Pharmacology and Vitreoretinal Surgery

Developments in Ophthalmology

Vol. 44

Series Editor

W. Behrens-Baumann Magdeburg

Pharmacology and Vitreoretinal Surgery

Volume Editor

Arnd Gandorfer Munich

30 figures, 13 in color, 2009

Basel · Freiburg · Paris · London · New York · Bangalore · Bangkok · Shanghai · Singapore · Tokyo · Sydney

Arnd Gandorfer
Vitreoretinal and Pathology Unit
Augenklinik der Ludwig-Maximilians-Universität
Mathildenstrasse 8
DE–80336 Munich (Germany)

The volume editor thanks ThromboGenics for the financial support to cover the costs of the color illustrations.

Library of Congress Cataloging-in-Publication Data

Pharmacology and vitreoretinal surgery / volume editor, Arnd Gandorfer.
p. ; cm. – (Developments in ophthalmology, ISSN 0250-3751 ; v. 44)
Includes bibliographical references and indexes.
ISBN 978-3-8055-9029-7 (hard cover : alk. paper)
1. Retina--Surgery. 2. Vitreous humor. 3. Enzymes--Therapeutic use. I. Gandorfer, Arnd.
[DNLM: 1. Vitreous Body--surgery. 2. Combined Modality Therapy. 3. Retina--surgery. 4. Retinal Diseases--surgery. 5. Vitrectomy--methods. 6. Vitreous Body--drug effects. W1 DE998NG v.44 2009 / WW 250 P536 2009]
RE551.P43 2009
617.7'3--dc22

2009009511

Bibliographic Indices. This publication is listed in bibliographic services, including Current Contents®.

www.karger.com
Printed in Switzerland on acid-free and non-aging paper (ISO 9706) by Reinhardt Druck, Basel
ISSN 0250–3751
ISBN 978–3–8055–9029–7
e-ISBN 978–3–8055–9030–3

Contents

List of Contributors

Paul N. Bishop
University of Manchester
Manchester Royal Eye Hospital
Oxford Road
Manchester M13 9WH (UK)

David G. Charteris
Moorfields Eye Hospital
City Road
London EC1V 2PD (UK)

Carlos E. Cury, Jr.
Vision Institute, Department of Ophthalmology
Federal University of Sao Paulo
Rua Presidente Coutinho 579
conj 501
Florianópolis
SC 88015–300 (Brazil)

Kimberly A. Drenser
Eye Research Institute, University Oakland
3535 West 13 Mile Road, Suite 344
Royal Oak, MI 48073 (USA)

Kirsten H. Eibl
University Eye Hospital LMU
Ludwig-Maximilians-University
Mathildenstr. 8
DE-80336 Munich (Germany)

Michel E. Farah
Vision Institute, Department of Ophthalmology
Federal University of Sao Paulo
Rua Presidente Coutinho 579
conj 501
Florianópolis
SC 88015–300 (Brazil)

Steven K. Fisher
Neuroscience Research Institute
University of California
Santa Barbara, CA 93106-5060 (USA)

Patrice E. Fort
Penn State Hershey Eye Center and
Department of Cellular and Molecular
Physiology
Penn State College of Medicine
500 University Drive, HU097
Hershey, PA 17033 (USA)

Arnd Gandorfer
Vitreoretinal and Pathology Unit
Augenklinik der Ludwig-Maximilians-
Universität
Mathildenstrasse 8
DE–80336 Munich (Germany)

Thomas W. Gardner
Penn State Hershey Eye Center and
Department of Cellular and Molecular
Physiology
Penn State College of Medicine
500 University Drive, HU097
Hershey, PA 17033 (USA)

David T. Goldenberg
3535 W. 13 Mile Rd., Suite 344
Royal Oak, MI 48073 (USA)

Hisanori Imai
Penn State Hershey Eye Center and
Department of Cellular and Molecular
Physiology
Penn State College of Medicine
500 University Drive, HU097
Hershey, PA 17033 (USA)

Motohiro Kamei
Department of Ophthalmology
Osaka University Medical School
2–2 Yamadaoka
Suita, 565-0871 (Japan)

Baruch Kuppermann
UC Irvine Medical Center
Ophthalmology Clinic
101 The City Drive South
Orange, CA 92697 (USA)

Geoffrey P. Lewis
Neuroscience Research Institute
University of California
Santa Barbara, CA 93106-5060 (USA)

Carsten H. Meyer
Department of Ophthalmology
University of Bonn
Ernst-Abbe-Strasse 2
DE–53127 Bonn (Germany)

Raja Narayanan
Smt Kanuri santhamma Vitreoretina center
L.V. Prasad Eye Institute
L.V. Prasad Marg, Banjara Hills
Hyderabad 500034 (India)

Eduardo B. Rodrigues
Vision Institute, Department of Ophthalmology
Federal University of Sao Paulo
Rua Presidente Coutinho 579
conj 501
Florianópolis
SC 88015–300 (Brazil)

Ravi S.J. Singh
Penn State Hershey Eye Center and
Department of Cellular and Molecular
Physiology
Penn State College of Medicine
500 University Drive, HU097
Hershey, PA 17033 (USA)

Yasuo Tano
Department of Ophthalmology
Osaka University Medical School
2–2 Yamadaoka
Suita, 565-0871 (Japan)

Michael T. Trese
3535 W. 13 Mile Rd., Suite 344
Royal Oak, MI 48073 (USA)

Louisa Wickham
Moorfields Eye Hospital
City Road
London EC1V 2PD (UK)

In memoriam of Professor Yasuo Tano, who suddenly passed away on 31 January 2009.

Preface

More than 35 years have passed since the advent of pars plana vitrectomy, and vitreoretinal surgery has developed to highly sophisticated techniques to treat retinal diseases. Removal of the vitreous gel and hemorrhage not only clears the optical axis of the eye, but enables the surgeon to approach the retina and the vitreoretinal interface directly, thereby relieving traction and removing pathological tissue, such as epiretinal membranes. Peeling off the internal limiting membrane has proven to be a safe and effective technique in macular surgery, resulting in macular hole closure. Twenty years ago, nobody would have imagined this. Today, final success rates beyond 90% can be achieved in macular and reattachment surgery.

However, there are limitations of current vitreoretinal surgery techniques, which are mechanically based. Removal of the vitreous is incomplete, especially at the vitreoretinal interface and at the vitreous base. This may lead to persistent or recurrent traction on the retina, resulting in retinal tear formation or reproliferation. More aggressive removal of the vitreous by mechanical means, however, carries the risk of damaging the retina. When epiretinal membranes are removed in PVR cases and in diabetic eyes with traction retinal detachment, gliotic scar tissue is removed but neural retina is not treated. Thus, despite anatomical reattachment, visual results are often disappointing.

Pharmacology-assisted vitreoretinal surgery can help to overcome these limitations. There are enzymes which cleave the vitreoretinal junction without damaging the retina, and those for liquefaction. Thereby, vitreolysis and induction of posterior vitreous detachment has become possible without the need for vitrectomy. Recent results from clinical trials show that up to 40% of eyes achieve release of traction without surgery.

Pharmacologic vitreolysis will change our current indications and concepts in treating retinal and macular diseases, and earlier intervention might save visual function before advanced stages have destroyed the retinal cytoarchitecture. In diabetic eyes, for example, enzymatic PVD induction at an early stage of the disease might inhibit fibrovascular and fibrocellular proliferations at the vitreoretinal interface, thereby preventing progression to proliferative disease.

Neuroprotective and antiproliferative agents may hold the promise of preserving neuronal function when the retina is detached or when PVR has developed. This again may change the time point of intervention from currently advanced stages to an earlier disease stage with less pathology. Fibrinolytic and antiproliferative agents help to treat disasters in ophthalmology such as massive submacular hemorrhage or retinal detachment in retinopathy of prematurity.

It is now time to combine pharmacological concepts and vitreoretinal surgery. World renowned experts and opinion leaders in their field have contributed their knowledge and skills to make this book the first summary on pharmacology-assisted vitreoretinal surgery. I am greatly indebted to the authors, and my hopes are that this book was a basis and motivation for clinicians and researchers who want to bring retinal surgery further by introducing pharmacology-assisted vitreoretinal surgery.

Arnd Gandorfer
Munich, 2009

Gandorfer A (ed): Pharmacology and Vitreoretinal Surgery.
Dev Ophthalmol. Basel, Karger, 2009, vol 44, pp 1–6

Objective of Pharmacologic Vitreolysis

Arnd Gandorfer

Vitreoretinal and Pathology Unit, University Eye Hospital Munich, Ludwig-Maximilians University, Munich, Germany

Abstract

The goal of pharmacologic vitreolysis is to cleave the vitreoretinal junction, thereby inducing posterior vitreous detachment (PVD), and to liquefy the vitreous gel. There are several reasons to pursue a pharmacologic approach: (1) Mechanical vitrectomy is incomplete. Both at the posterior pole and in the retinal periphery, remnants of the cortical vitreous are left behind at the internal limiting membrane of the retina, causing vitreoretinal traction and (re)proliferation of cells, and leading to surgical failure. (2) Pharmacologic vitreolysis offers complete PVD without mechanical manipulation at the vitreoretinal interface, such as ILM peeling, thereby minimizing the risk of iatrogenic damage to the macula. (3) An intravitreal injection resulting in complete PVD is a less traumatic approach than vitrectomy, and it might be beneficial as prophylactic treatment regime in retinal diseases characterized by fibrocellular and fibrovascular proliferation at the vitreoretinal interface, such as diabetic macular edema and proliferative retinopathy, in order to prevent advanced stages of disease. (4) Cleaving the cortical hyaloid completely from the retina changes the molecular flux across the vitreoretinal interface and improves oxygen supply to the retina, a major mechanism of action which might significantly interfere with biochemical pathways of retinal hypoxia, leading to an overexpression of vasoactive substances such as vascular endothelial growth factor.

The vitreous, and in particular the vitreoretinal interface, plays an important role in the pathogenesis of many retinal disorders. An abnormal interface has been implicated in vitreomacular traction syndrome, macular holes, diabetic macular edema, proliferative diabetic retinopathy, retinal detachment, and proliferative vitreoretinopathy [1]. A common finding in these entities is a vitreous which is firmly attached to the retina, thereby preventing complete posterior vitreous detachment (PVD) [2, 3]. Even in cases of vitreous liquefaction, the vitreous cortex remains attached to the retina, forming so-called 'vitreoschisis', which acts as a scaffold for fibrocellular and fibrovascular proliferations [4–6]. This may increase traction on the retina leading to significant patient morbidity and surgical failure.

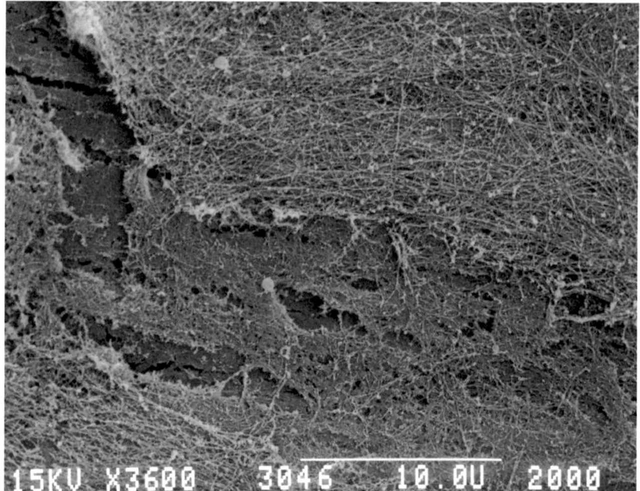

Fig. 1. Scanning electron micrograph of the vitreoretinal interface showing remnants of cortical vitreous attached at the ILM following mechanical separation of the vitreous from the retina by using suction with the vitrectomy probe.

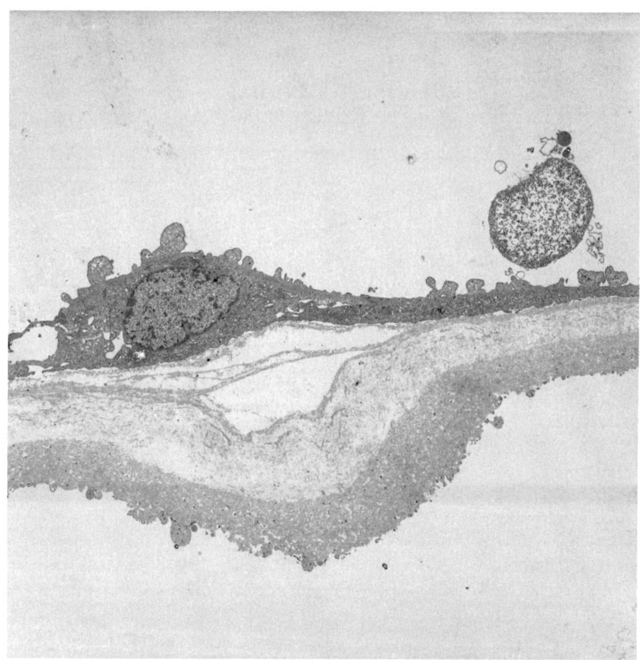

Fig. 2. ILM specimen removed from an eye with vitreomacular traction syndrome. On the vitreal side of the ILM, there is cortical vitreous and fibrocellular proliferation which is removed completely together with the ILM.

Vitrectomy is the treatment modality of choice in these patients, and complete removal of the cortical hyaloid is critical for success of the operation [2, 7]. However, mechanical separation of the vitreous from the retina is incomplete, as cortical vitreous fibrils are left behind on the internal limiting membrane (ILM) of the retina (fig. 1) [6]. Removal of the ILM creates a new vitreomacular interface by eliminating the vitreous cortex and proliferative tissue (fig. 2). However, ILM removal requires direct

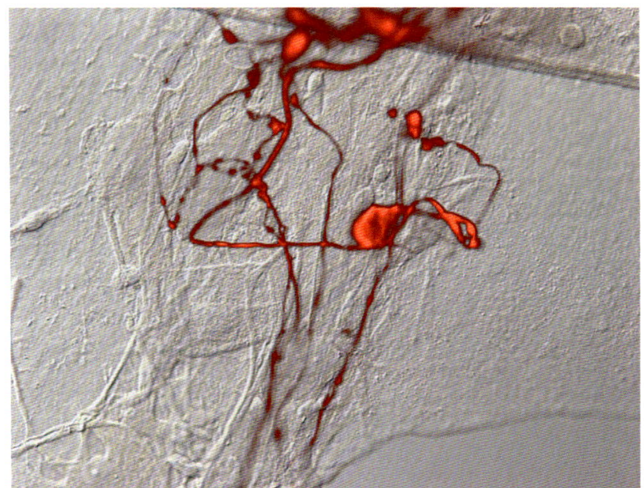

Fig. 3. ILM specimen removed from an eye with macular hole. Immunocytochemistry using antibody against neurofilament, a specific marker of retinal ganglion cells. Note the ganglion cell and its axon removed together with the ILM.

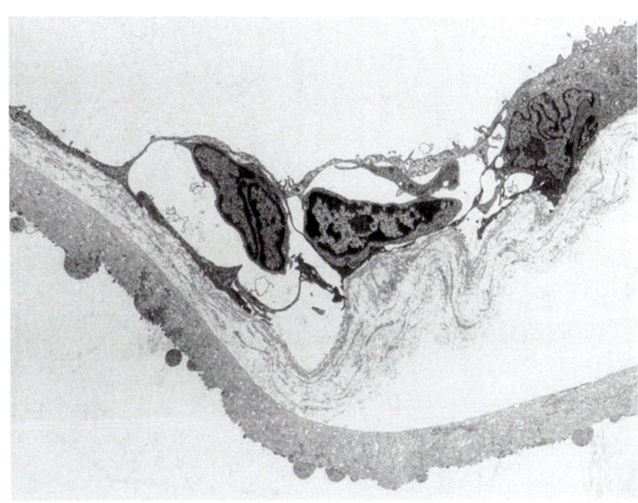

Fig. 4. Pharmacologic vitreolysis causing complete PVD cleaves the vitreoretinal junction at the vitreal side of the ILM, thereby removing cortical vitreous together with cellular proliferations exerting traction.

manipulation of the macula which can potentially result in retinal damage, leading to functional defects, such as central or paracentral scotomata [8, 9]. Ultrastructural analysis of peeled ILM specimens has shown damage to Müller cells and avulsion of macular ganglion cells (fig. 3) [8, 10]. Although today, ILM peeling appears safe in clinical terms, it may not be the best option ever in terms of safety and visual benefit.

Cleaving the vitreoretinal interface at the vitreal side of the ILM would be a logical approach to relieve traction, as the ILM itself cannot exert traction (fig. 4). Traction is generated by contractile cells at the vitreoretinal interface [11]. In most instances, these cells proliferate on a layer of vitreous cortex which remained attached at the ILM after incomplete PVD [5, 12, 13]. From an ultrastructural point of view, separation of

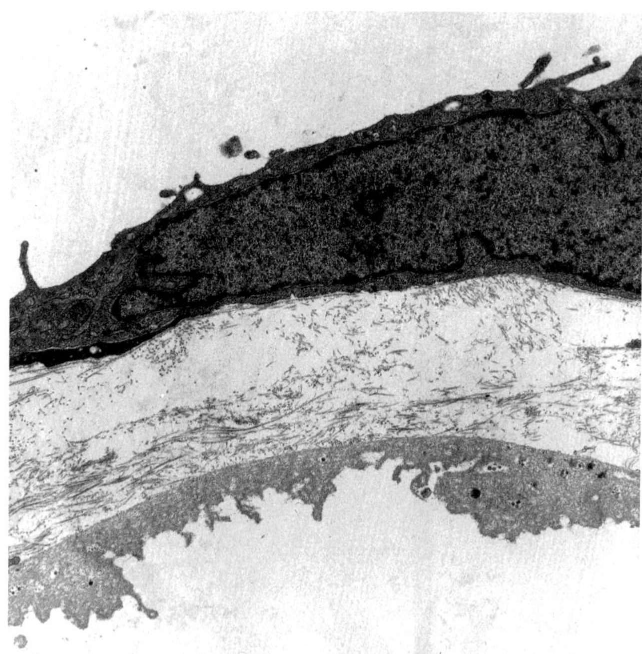

Fig. 5. The cortical vitreous forms a scaffold for fibrocellular and fibrovascular proliferations, especially in diabetic eyes. Clinically, this may be associated with a thickened and taut premacular hyaloid.

the cortical vitreous completely from the ILM would eliminate the cellular proliferations, thereby relieving the traction they generate.

Complete separation of the vitreous from the retina by enzymatic means would remove the pathologic tissue from the retina, and eliminate the scaffold for cellular re-proliferation [14–17]. Especially diabetic eyes are characterized by a firm vitreoretinal attachment and a thickened and taut vitreous cortex which frequently contains cellular proliferations (fig. 5) [3, 12]. Given that some cells have been shown to produce vasoactive and vasoproliferative substances, such as vascular endothelial growth factor and others, complete removal of the vitreous cortex with embedded cells could lead back to a more physiological state of the vitreomacular interface in patients with diabetic macular edema [18].

Moreover, a clean ILM without remnants of the cortical vitreous would hypothetically prevent the growth of fibrovascular membranes out of the retina, as these newly formed vessels commonly proliferate within and along the cortical hyaloid [19]. Given that a complete PVD was achieved by a relatively atraumatic intravitreal injection and not by an operation such as vitrectomy, pharmacologic vitreolysis could be discussed in select cases as a prophylactic treatment regime to prevent proliferative diabetic retinopathy [2].

Another mechanism of action of enzymatically induced PVD and vitreous liquefaction is an alteration of molecular flux into the vitreous cavity, out of the vitreous cavity, and across the vitreoretinal interface [20]. It has been know since the 1980s that oxygen levels within the vitreous cavity are higher in vitrectomized eyes compared to nonvitrectomized eyes [21]. Recently, it has been demonstrated that the oxygen supply of the retina can be modified by a microplasmin-induced PVD and vitreous liquefaction [20]. It is of note that liquefaction alone did not have this effect. Obviously, the vitreous cortex acts as a barrier for molecules when crossing the vitreoretinal border, and this may not only be the case for oxygen but also when other molecules are produced within the eye or even when are injected intravitreally [22]. A better oxygen supply of the retina addresses both diseases with hypoxia such as diabetic retinopathy and retinal vein occlusion, both complicated by macular edema and neovascularization, and those in which anti-VEGF treatment is performed in order to decrease intravitreal VEGF levels such as age-related macular degeneration. Interestingly, eyes with wet AMD also show a significantly higher percentage of attached vitreous at the macula compared to non-/dry AMD eyes. Separation of the posterior hyaloid from the retina and liquefaction alters the molecular flux across the vitreoretinal interface, and it might be expected that larger molecules are even more affected by the barrier function of the cortical vitreous than small molecules such as oxygen.

The summary of potential objectives of pharmacologic vitreolysis includes complete vitreoretinal separation resulting in a clean ILM, less manipulation at the vitreoretinal interface compared to current vitrectomy techniques, a well-defined cleavage plane at the vitreal side of the ILM to remove fibrocellular traction, better oxygen supply due to an alteration of the molecular flux, and a less traumatic and maybe prophylactic treatment regime in select cases in the future [2, 23, 24].

References

1 Sebag J: The Vitreous: Structure, Function, Pathobiology. New York, Springer, 1989.

2 Gandorfer A: The need for pharmacology in vitreoretinal surgery SOE Lecture 2007. Klin Monatsbl Augenheilkd 2007;224:900–904.

3 Sebag J: Diabetic vitreopathy. Ophthalmology 1996; 103:205–206.

4 Schwartz SD, Alexander R, Hiscott P, Gregor ZJ: Recognition of vitreoschisis in proliferative diabetic retinopathy: a useful landmark in vitrectomy for diabetic traction retinal detachment. Ophthalmology 1996;103:323–328.

5 Gandorfer A, Rohleder M, Kampik A: Epiretinal pathology of vitreomacular traction syndrome. Br J Ophthalmol 2002;86:902–909.

6 Gandorfer A, Ulbig M, Kampik A: Plasmin-assisted vitrectomy eliminates cortical vitreous remnants. Eye 2002;16:95–97.

7 Gandorfer A: Vitrektomie. Z Prakt Augenheilk 2008; 29:249–261.

8 Gandorfer A, Rohleder M, Charteris DG, Sethi C, Kampik A, Luthert P: Staining and peeling of the internal limiting membrane in the cat eye. Curr Eye Res 2005;30:977–987.

9 Haritoglou C, Gass CA, Kristin N, Kampik A: Paracentral scotomata: a new finding after vitrectomy for idiopathic macular hole. Br J Ophthalmol 2001;85:231–233.

10 Gandorfer A, Scheler R, Schumann R, Haritoglou C, Kampik A: Interference microscopy delineates cellular proliferations on flat mounted internal limiting membrane specimens. Br J Ophthalmol 2009; 93:120–122.

11 Kampik A, Green WR, Michels RG, Nase PK: Ultrastructural features of progressive idiopathic epiretinal membrane removed by vitreous surgery. Am J Ophthalmol 1980;90:797–809.

12 Gandorfer A, Rohleder M, Grosselfinger S, Haritoglou C, Ulbig M, Kampik A: Epiretinal pathology of diffuse diabetic macular edema associated with vitreomacular traction. Am J Ophthalmol 2005; 139:638–652.

13 Gandorfer A, Rohleder M, Charteris D, Kampik A, Luthert P: Ultrastructure of vitreomacular traction syndrome associated with persistent hyaloid artery. Eye 2005;19:333–336.

14 Gandorfer A: Enzymatic vitreous disruption. Eye 2008;22:1273–1277.

15 Gandorfer A, Rohleder M, Sethi C, Eckle D, Welge-Lussen U, Kampik A, Luthert P, Charteris D: Posterior vitreous detachment induced by microplasmin. Invest Ophthalmol Vis Sci 2004;45:641–647.

16 Gandorfer A, Priglinger S, Schebitz K, Hoops J, Ulbig M, Ruckhofer J, Grabner G, Kampik A: Vitreoretinal morphology of plasmin-treated human eyes. Am J Ophthalmol 2002;133:156–159.

17 Gandorfer A, Putz E, Welge-Lussen U, Gruterich M, Ulbig M, Kampik A: Ultrastructure of the vitreoretinal interface following plasmin assisted vitrectomy. Br J Ophthalmol 2001;85:6–10.

18 Gandorfer A: Pharmacologic vitreolysis. Dev Ophthalmol 2007;39:149–156.

19 Faulborn J, Bowald S: Microproliferations in proliferative diabetic retinopathy and their relationship to the vitreous: corresponding light and electron microscopic studies. Graefes Arch Clin Exp Ophthalmol 1985;223:130–138.

20 Quiram PA, Leverenz VR, Baker RM, Dang L, Giblin FJ, Trese MT: Microplasmin-induced posterior vitreous detachment affects vitreous oxygen levels. Retina 2007;27:1090–1096.

21 Stefansson E, Novack RL, Hatchell DL: Vitrectomy prevents retinal hypoxia in branch retinal vein occlusion. Invest Ophthalmol Vis Sci 1990;31:284–289.

22 Sebag J, Ansari RR, Suh KI: Pharmacologic vitreolysis with microplasmin increases vitreous diffusion coefficients. Graefes Arch Clin Exp Ophthalmol 2007; 245:576–580.

23 Trese M: Enzymatic-assisted vitrectomy. Eye 2002; 16:365–368.

24 Sebag J: Is pharmacologic vitreolysis brewing? Retina 2002;22:1–3.

Arnd Gandorfer, MD, PD, FEBO
Vitreoretinal and Pathology Unit, Augenklinik der Ludwig-Maximilians-Universität
Mathildenstrasse 8
DE–80336 Munich (Germany)
Tel. +49 89 5160 3800, Fax +49 89 5160 4778, E-Mail arnd.gandorfer@med.uni-muenchen.de

Gandorfer A (ed): Pharmacology and Vitreoretinal Surgery.
Dev Ophthalmol. Basel, Karger, 2009, vol 44, pp 7–19

Vitreous as a Substrate for Vitreolysis

Paul N. Bishop

University of Manchester and Manchester Biomedical Research Centre, Manchester Royal Eye Hospital, Manchester, UK

Abstract

There is an increasing interest in the concept of using enzymes to induce vitreoretinal separation and/or modulate the structure of the vitreous as an aid to, or replacement for, mechanical vitrectomy. In order to understand how these enzymes work and to develop new and better therapeutics, knowledge of vitreous structure at a molecular and supramolecular level is key. The vitreous like many extracellular matrices has a composite structure with a network of collagen fibrils, that is essential for the gel structure, and a glycosaminoglycan network composed predominantly of hyaluronan that fills the spaces between the collagen fibrils and helps to inflate the matrix. The collagen fibrils are very resistant to proteolytic degradation so these do not make a good target for pharmacological vitreolysis, but degradation of the hyaluronan facilitates the dispersal of vitreous haemorrhage. The molecular mechanisms underlying postbasal vitreoretinal adhesion are poorly understood but it is likely that intermediary molecules that link cortical vitreous collagen fibrils to the inner limiting lamina play a major role. Furthermore, it is apparent that proteolytic enzymes such as plasmin fragment these linking molecules and thereby facilitate vitreoretinal disinsertion.

There has been considerable interest in recent years in the use of pharmacological agents to facilitate, or indeed replace mechanical vitrectomy. The main research focus has been on enzymes that cleave the vitreoretinal interface, modify vitreous structure, or a combination of the two. These agents have the potential to facilitate the treatment of conditions where there is vitreoretinal traction or to create a posterior vitreous detachment (PVD) to protect against pathologies such as proliferative diabetic retinopathy. Alternatively, agents that modify gel structure may be used, for example, to facilitate the clearance of blood from the vitreous cavity.

The use of enzymes in the ocular cavity has a long history with, for example, chymotrypsin being used to disrupt the zonules during intracapsular cataract surgery. Interestingly this did not appear to cause significant collateral damage to ocular tissues. In the 1970s, experiments were undertaken to determine whether bacterial collagenases could be used for vitreolysis, but these were abandoned when it became apparent that intravitreal collagenase caused collateral damage to the retinal vasculature. These early experiments did however focus attention on one of the key issues in this research field, i.e. to identify pharmacological agents that assist or replace mechanical vitrectomy, but do not cause collateral damage to other ocular structures. To achieve this aim it is helpful to understand, at a molecular level, how the vitreous and vitreoretinal interface are organised, so that specific molecules or molecular interactions can be targeted with appropriate pharmacological agents.

Structural Molecules and Macromolecular Organisation of the Vitreous Humour

Structural Macromolecules

The vitreous humour is a unique extracellular matrix that normally only contains only a few cells, mainly hyalocytes, that are localised in its periphery. Extracellular matrices are formed and maintained by network-forming macromolecules with complementary properties. Proteins such as collagens endow tissues with shape, strength, flexibility and resistance to tractional forces. Carbohydrates, particularly glycosaminoglycans (GAGs), fill the spaces between the collagens, inflate the tissue and resist compressive forces. GAGs are formed from long chains of negatively charged sugar units that attract counter-ions and water and thereby provide a swelling pressure that spaces apart and stabilises the collagen network. In the vitreous the major components of this composite structure are fibrillar collagens and the GAG hyaluronan (fig. 1).

Collagens and Associated Molecules

The vitreous gel contains a low concentration of collagen that is estimated to be 300 μg/ml in human eyes [1]. Despite this low concentration, it is the collagen fibrils that impart gel-like properties to the vitreous. In 1982, Balazs and Denlinger [2] showed that the total amount of vitreous collagen does not change throughout life so the concentration decreases as the eye grows; they therefore suggested that there is no postnatal synthesis of vitreous collagen. However, more recent research suggests that there is some postnatal synthesis of collagen, although probably at a low- level [1, 3–5].

Collagen molecules are formed by three polypeptide chains (α-chains) that fold in a characteristic triple-helical configuration [1]. This triple-helical structure requires that every third amino acid is a glycine residue, so collagen α-chains have $(Gly\text{-}X\text{-}Y)_n$ triplet amino acid repeats where X and Y can be any amino acids, but are frequently proline and hydroxyproline. Hydroxyproline is important in stabilising the structure of the collagen triple helix by forming extra hydrogen bonds. The collagen molecules assemble into fibrils and the amino acids lysine and hydroxylysine then participate in crosslink formation within the fibrils. The net result is the formation of very strong

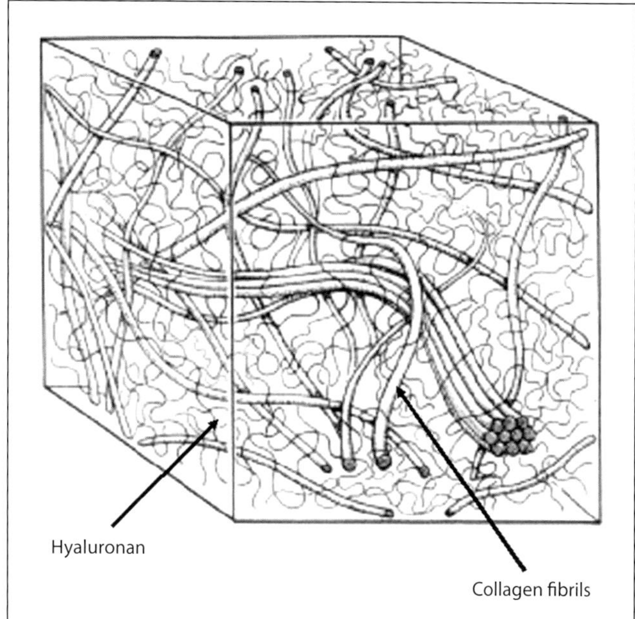

Fig. 1. Schematic representation of the collagen and hyaluronan networks of the vitreous. The fibrillar collagen network maintains the gel state and provides the tensile strength whereas the hyaluronan network fills the spaces between these collagen fibrils and provides a swelling pressure to inflate the gel.

rope-like structures that are resistant to the actions of many proteolytic enzymes. Enzymes that are known to breakdown collagen include bacterial collagenase, some metalloproteinases and cathespin K.

There are at least 27 different types of collagen molecules (i.e. collagen types I–XXVII) and these typically assemble into fibrils or sheet-like structures. In the vitreous humour nearly all of the collagen is in the form of thin, uniform, heterotypic (of mixed composition) fibrils containing collagen types II, IX and V/XI [1, 6]. It is of note that cartilage collagen fibrils have a very similar composition. There are also type VI collagen microfibrils in the vitreous, but they are in low abundance and their significance is unknown [1].

The predominant component of the heterotypic collagen fibrils (60–75%) is type II collagen [1]. Molecules of this collagen are composed of three identical α-chains, i.e. its composition is $\alpha 1(II)_3$. When fibrillar collagens such as type II collagen molecules are secreted into the extracellular environment they are in a soluble precursor form, the procollagen. Procollagen has terminal extensions called the N- and C-propeptides. Once within the extracellular environment, the procollagen is 'processed' by specific enzymes called N-proteinase and C-proteinase (or BMP-1) that cleave the N- and C-propeptides respectively, leaving small non-collagenous telopeptides at each end of the triple-helical region. This process reduces the solubility of the collagen molecules and allows them to participate in fibril formation.

Type IX collagen has been estimated to represent up to 25% of the collagen in the heterotypic collagen fibrils of vitreous [1]. It covalently links to the surface of these

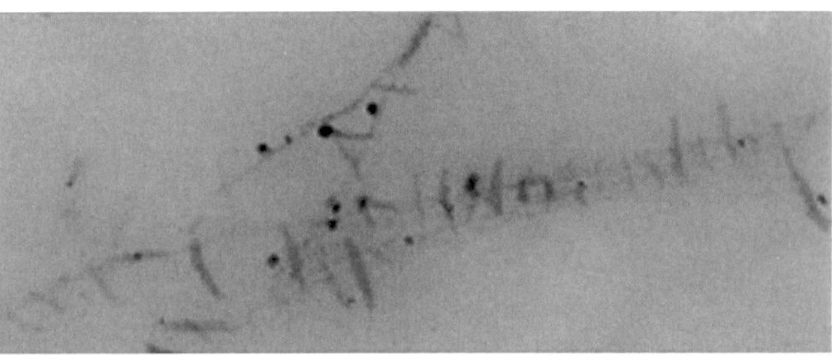

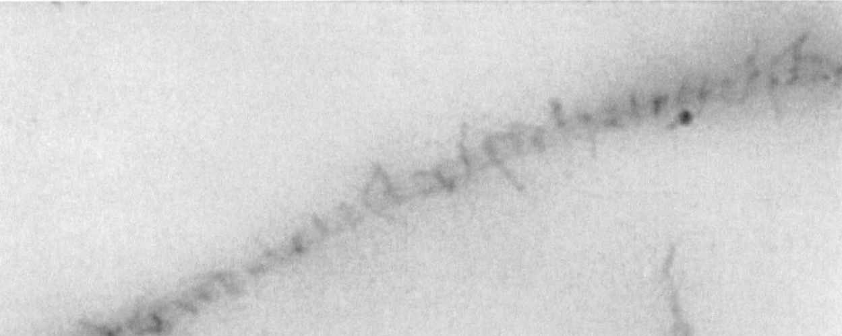

Fig. 2. Electron micrographs of vitreous collagen fibrils stained with uranyl acetate (labelling collagen fibrils) and cupromeronic blue (labelling the chondroitin sulphate chains of the type IX collagen). The collagen fibrils are organised into bundles and within these bundles the type IX collagen chondroitin sulphate chains appear to bridge between adjacent fibrils in a ladder-like configuration. Therefore, these chondroitin sulphate chains may both space apart and interconnect the collagen fibrils.

fibrils in a regular periodicity (the D-period). It is a member of the family of fibril associated collagens with interrupted triple helices (FACIT), and a common feature of these collagens is a multiple-domain structure with triple-helical domains (COL domains) that are separated or interrupted by short non-triple-helical domains (NC domains). Type IX collagen is a heterotrimer made of 3 distinct polypeptide chains i.e. its chain composition is α1(IX)α2(IX)α3(IX) and these assemble to produce three collagenous domains (COL1, COL2 and COL3) interspersed between four non-collagenous domains (NC1, NC2, NC3 and NC4). In vitreous, the type IX collagen is a proteoglycan and is synthesised with a single covalently-linked chondroitin sulphate GAG chain. These chondroitin sulphate chains can be visualised by electron microscopy after labelling with the cationic dye cupromeronic blue (fig. 2).

The third collagen in the heterotypic collagen fibrils is another fibrillar collagen called type V/XI collagen and this represents 10–25% of the collagen in the fibrils [1]. It co-assembles with the type II collagen to form the core of the fibrils. This collagen contains α1(XI) and α2(V) chains, probably with a stoichiometry of $\alpha 1(XI)_2\alpha 2(V)$. The N-terminal domain of type XI collagen is incompletely removed by processing and may play a role in regulating the fibril diameter.

The heterotypic collagen fibrils of the vitreous are coated with a molecule called opticin [7, 8]. This is a member of the extracellular matrix small leucine-rich repeat protein/proteoglycan (SLRP) family. Most members of this family are proteoglycans, but opticin is instead glycosylated via a cluster of O-linked oligosaccharides. Opticin exists as a dimer in solution as a result of interactions between the LRR domains [9]. Opticin binds to GAGs including heparan sulphate and chondroitin sulphate and may thereby play a role in stabilising the vitreous gel structure and in adhesion at the vitreoretinal interface [10].

GAGs and Proteoglycans

GAGs are composed of long chains of repeating disaccharide units. All GAGs except hyaluronan are synthesised attached to a protein core (thereby forming proteoglycans) and undergo varying degrees of modification during their synthesis including sulphation, epimerisation and acetylation/deacetylation. The predominant GAG in vitreous is hyaluronan (>90%), but it also contains small amounts of chondroitin sulphate that may be important to its structure.

Hyaluronan is formed from repeating disaccharide units [β1–4 glucuronic acid β1–3 N-acetylglucosamine]$_n$. Chains of hyaluronan can be very long and form networks through entanglement. It is distinguished from the other GAGs in that it is not synthesised covalently linked to a core protein and it is not sulphated. Hyaluronan is not uniformly distributed within the vitreous and the highest concentration is found in the posterior vitreous cortex. In the adult human vitreous the hyaluronan concentration has been estimated to be between 65 and 400 μg/ml and the average molecular weight to be 2–4 million [1].

The repeating disaccharide unit of chondroitin sulphate chains is [β1–4 glucuronic acid β1–3 N-acetylgalactosamine]$_n$. Sulphation is added to these chains are various points. The vitreous has been shown to contain two chondroitin sulphate proteoglycans, type IX collagen (see above) and versican [11]. Versican is a large proteoglycan with a central domain that carries multiple chondroitin sulphate chains. It possesses an N-terminal hyaluronan-binding domain and the binding to hyaluronan is stabilised by link protein, which has also identified in the vitreous. Therefore, it is likely that the versican in vitreous is linked to the hyaluronan network.

GAGs can be degraded by specific enzymes. Hyaluronidases, also called hyaluronan lyases, are enzymes that degrade hyaluronan. They are derived from different sources and have different specificities. For example, testicular hyaluronidase (bovine or ovine) degrades chondroitin and dermatan sulphate as well as hyaluronan, and some preparations have significant protease activity. By contrast, the hyaluronidase derived from *Streptomyces* is specific for hyaluronan. The enzyme chondroitin ABC lyase degrades chondroitin and dermatan sulphate, and it addition has some activity against hyaluronan.

Supramolecular Organisation

The gel state of the vitreous is maintained by the network of heterotypic collagen fibrils as their removal, for example by centrifugation, converts the vitreous into a viscous liquid. This network of collagen fibrils can be observed by electron microscopy.

Transmission electron microscopy reveals that these collagen fibrils are unusually thin, approximately 15 nm in diameter, and unbranched [6]. Analyses by freeze-etch rotary shadowing electron microscopy showed that the collagen fibrils are arranged in bundles within the vitreous and form an extended interconnected network by branching between these bundles [12]. In the young vitreous, the collagen fibrils within these bundles appear to run closely together in parallel, but are not fused (figs. 2, 5). Electron microscopic studies, using cationic dyes, suggest that the chondroitin sulphate chains of type IX collagen play a role in both connecting together and spacing apart the collagen fibrils within these bundles [13].

Hyaluronan provides a swelling pressure to the collagen network and hence the vitreous gel. However, the enzymatic degradation of the hyaluronan in isolated bovine vitreous gels resulted in some gel shrinkage, but not complete collapse; so hyaluronan is not a prerequisite for the bovine vitreous gel structure [14]. Further studies demonstrated that hyaluronan weakly associates with vitreous collagen fibrils and the hyaluronan network 'stiffens' the collagenous network, as hyaluronidase digestion resulted in deflation and relaxation of the collagen network [12]. Taken together these data suggest that the gel structure can exist without hyaluronan, but that it is deflated. Therefore, hyaluronidases are unlikely to cause extensive liquefaction of the vitreous. However, hyaluronan contributes towards the trapping of cells in the vitreous cavity, for example following vitreous haemorrhage, and the enzymatic depolymerisation of hyaluronan facilitates the dispersion of trapped cells.

Vitreoretinal Adhesion

Morphological analyses reveal that the mechanisms underlying vitreoretinal adhesion differ within and posterior to the vitreous base. Within the vitreous base there is effectively an unbreakable adhesion and this is a result of the vitreous collagen fibrils passing through defects in the inner limiting lamina (ILL) and intertwining with collagen on the cellular side of this basement membrane [3]. Indeed the interlinking of newly synthesised collagen in the peripheral retina that has penetrated the ILL with the pre-existing cortical vitreous collagen may account for the posterior migration of the posterior border of the vitreous base with aging [3].

Posterior to the vitreous base the cortical vitreous collagen fibrils do not generally insert into or pass through the post-oral ILL [1]. However, there may be focal areas where direct collagenous connections are formed through the ILL such as in strong paravascular vitreoretinal adhesions [15]. Nonetheless, for most of the postoral vitreoretinal interface, the cortical vitreous collagen fibrils run parallel with the ILL and the basis of adhesion is likely to be interactions between molecules on the surface of the cortical collagen fibrils and inner surface of the ILL (fig. 3). The molecular basis of these adhesions is poorly understood. The ILL is a basement membrane that contains molecules including type IV collagen, laminins and

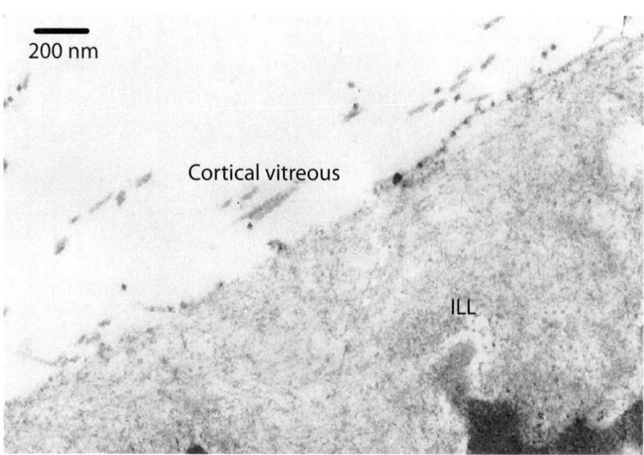

Fig. 3. At the post-basal vitreo retinal interface, the collagen fibrils do not generally insert directly into the ILL, but instead run parallel with its inner surface. This morphology suggests that there are intermediary molecules linking the cortical vitreous collagen to the ILL and that these provide the molecular basis of vitreoretinal adhesion.

heparan sulphate proteoglycans such as type XVIII collagen. It has been shown that collagen type XVIII contributes towards vitreoretinal adhesion in the mouse eye, as a proportion of eyes from type XVIII collagen knockout mice had vitreoretinal disinsertion [16]; this may have been due to abnormal development at the vitreoretinal interface or due to type XVIII collagen being directly involved in vitreoretinal adhesion. Type XVIII collagen is a heparan sulphate proteoglycan and opticin binds to heparan sulphate [10]. Therefore, it is possible that opticin on the surface of cortical vitreous collagen fibrils binds to heparan sulphate proteoglycans of the ILL including type XVIII collagen, thus providing a molecular basis for vitreoretinal adhesion. However, opticin knockout mice do not appear to have spontaneous vitreoretinal disinsertion [unpubl. obs.], suggesting that in the mouse eye this interaction is not critical for vitreoretinal adhesion and that other molecular interactions are also involved.

Vitreous Liquefaction and Posterior Vitreous Detachment

The human vitreous humour undergoes an inevitable process of liquefaction (or syneresis) with aging so that by the age of 80–90 years more than half of the vitreous is liquefied [1]. Ultrastructural studies have shown that collagen fibrils aggregate with aging into macroscopic strands within the vitreous gel [17]. Age-related vitreous liquefaction may be caused by this gradual and progressive aggregation of the collagen fibrils resulting in a redistribution of the collagen fibrils leaving areas devoid of collagen fibrils and thereby converted into a liquid, and the collagen aggregates concentrated in the residual gel (fig. 4, 5). An alternative hypothesis is that age-related liquefaction is a result of destruction of vitreous collagen fibrils through enzymatic activity [18].

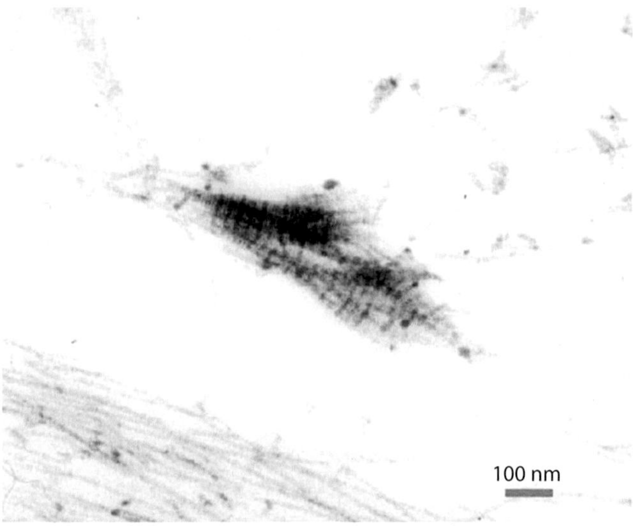

Fig. 4. During aging the vitreous collagen fibrils progressively aggregate together. This may lead to a redistribution of the collagen, with the aggregates collagen fibrils being concentrated in the residual gel structure and other areas becoming devoid of collagen resulting in their liquefaction.

There is an age-related loss of type IX collagen from the surface of the heterotypic collagen fibrils of the human eye [13]. Indeed, the half-life for the type IX collagen was found to be just 11 years of age. The chondroitin sulphate chains of type IX collagen appear to space collagen fibrils apart within the bundles that have been observed in the vitreous, so the age-related depletion of type IX collagen from the fibril surfaces will result in a loss of the spacers between the collagen fibrils. Once they come into contact, the stickiness of the fibrillar collagens will result in a tendency towards fusion. These conclusions are supported by a study showing that digestion of the vitreous with chondroitin ABC lyase, an enzyme that degrades the CS chains of type IX collagen, resulted in the aggregation of collagen fibrils [19]. Therefore, the aggregation of collagen that may result in vitreous liquefaction is caused by a loss of type IX collagen.

PVD is the process whereby the cortical vitreous gel splits away from the inner limiting lamina on the inner surface of the retina as far anteriorly as the posterior border of the vitreous base. It occurs spontaneously in approximately 25% of the population during their lifetime. PVD can be localised, partial or 'complete' (i.e. up to the posterior border of the vitreous base). PVD results from a combination of vitreous liquefaction and weakening of post-basal vitreoretinal adhesion. During PVD separation of the basal vitreous from the peripheral retina and ciliary body does not occur, because of the very strong adhesion in this region.

Whilst in a majority of subjects PVD occurs without major complications, in some it has sight-threatening complications. The concept of anomalous PVD was introduced by Sebag [20] to describe the situation where the extent of liquefaction exceeds weakening of vitreoretinal adhesion. This results in excessive tractional forces being exerted upon the retina during PVD and/or incomplete separation thereby leading

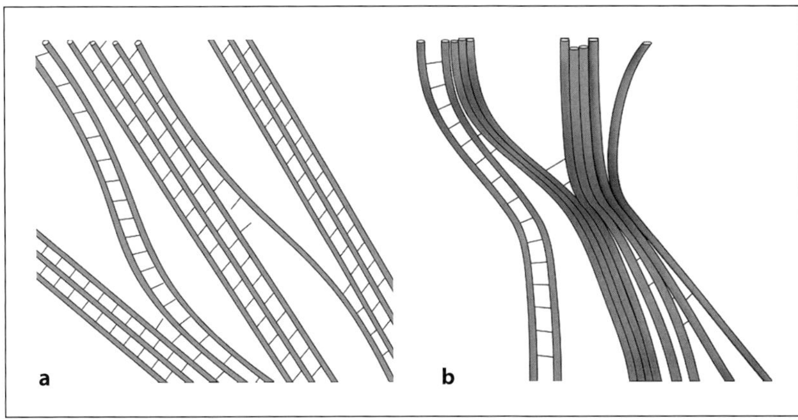

Fig. 5. The collagen fibrillar network of the vitreous and ageing changes. **a** The collagen fibrils (thick grey lines) form an extended network by being organised into small bundles that are interconnected by collagen fibrils running from one bundle to another. Within each bundle the collagen fibrils are both connected together and spaced apart by the CS chains of type IX collagen (thin black lines). **b** With ageing there is a loss of type IX collagen from the fibril surfaces. The loss of the type IX collagen CS chains from the fibril surfaces combined with an increased surface exposure of sticky type II collagen results in collagen fibrillar aggregation.

to complications including haemorrhage, retinal tears and detachment, macular hole formation and vitreomacular traction syndrome.

Enzymatic Vitreolysis

There has been increasing interest over the past few years in the use of intravitreal enzymes to cleave the vitreoretinal interface and/or altering the structure of the vitreous. When intravitreal enzymes are used the dosing and duration of action in the vitreous cavity are important parameters to consider as too little enzyme will be ineffective and too much may cause collateral damage to other ocular tissues, particularly the retina. Different animal models have been used in recent studies, along with enucleated human eyes. Some of these studies have progressed to clinical trials.

Plasmin, Microplasmin and Tissue Plasminogen Activator

At the present time the most promising enzymes for inducing vitreoretinal separation are plasmin and microplasmin (Thrombogenetics Ltd., Dublin). Microplasmin is a recombinant protein that contains the catalytic domain of human plasmin. Plasmin and microplasmin are non-specific serine proteases that cleave a variety of glycoproteins including fibronectin, laminin, fibrin and thrombospondin. Whilst laminin and fibronectin have been identified at the vitreoretinal interface and it has been shown that they are cleaved during plasmin-assisted vitrectomy [21], there is not direct evidence that these glycoproteins contribute towards vitreoretinal adhesion. It has been proposed that plasmin acts indirectly by cleaving endogenous (inactive) pro-metalloproteinases (pro-MMPs) and thereby converting them into the active MMP: Takano et al. [22] showed that intravitreal plasmin injection resulted in the activation of endogenous pro-MMP2. Indeed it has been shown that vitreous contains sufficient endogenous pro-MMP2 to cause significant destruction of the gel when activated [23].

In an investigation using post-mortem human eyes, plasmin was shown to induce complete vitreoretinal disinsertion detaching the cortical vitreous collagen from the inner limiting membrane as far as the posterior border of the vitreous base [24]. Scanning electron microscopy of the inner limiting membrane showed only sparse vitreous collagen fibrils remaining. Microplasmin increases vitreous diffusion coefficients, implying that it alters vitreous structure as well as inducing vitreoretinal disinsertion [25].

It was first reported by Verstraeten et al. [26] that when 1 U of plasmin was injected in rabbit eyes vitreous detachment was facilitated during vitrectomy. This group observed a transient decrease in electroretinogram b-wave amplitude, but no other adverse effects. Several studies supported these initial findings and subsequently clinical studies were performed autologous plasmin. Plasmin-assisted vitrectomy was performed is case series with traumatic macular holes, diabetic retinopathy and adult stage 3 macular holes and in all of these studies autologous plasmin was reported to facilitate the vitrectomy [27–29]. Further studies supported the observations that autologous plasmin when delivered before a mechanical vitrectomy was safe and facilitated the procedure [30, 31].

Gandorfer et al. [32] evaluated the effects of 62.5–188 μg intravitreal microplasmin on post-mortem human eyes and 14.5–25 μg on feline eyes in vivo. 125–188 μg of microplasmin induced PVD in human eyes and 25 μg in the feline eyes and an ILL that was free of cortical vitreous collagen fibrils by scanning electron microscopy. There was no evidence of cellular damage to the feline retinas. Microplasmin was delivered by intravitreal injection into rabbit eyes in a dose range of 12.5–250 μg [33]. A transient inflammatory reaction and reduction in ERG was observed, with the effects on ERG being less at doses of less than 250 μg. Doses of 125 μg or higher induced PVD and an ILL devoid of collagen fibrils when examined by scanning electron microscopy. Thrombogenetics have instituted clinical trials to evaluate the safety and efficacy of microplasmin used in combination with mechanical vitrectomy (for vitreomacular traction), or used alone (for diabetic macular oedema). Trial reports suggest safety and efficacy, but these have yet to be published.

A further approach that has been used is to inject tissue plasminogen activator into the vitreous cavity. This converts endogenous vitreous plasminogen into plasmin. When 25 μg was injected shortly prior to mechanical vitrectomy it was reported to facilitate the procedure [34]. However, a subsequent study did not support these observations [35].

Dispase

Dispase is a neutral protease derived from *Bacillus polymyxa* that preferentially cleaves fibronectin and type IV collagen [36]. Initial studies using procine and human cadaver eyes [37] and then in vivo experiments using porcine eyes [38] suggested that intravitreal dispase could induce or facilitate PVD without causing significant collateral damage to other ocular tissues. However, subsequent studies found evidence of retinal haemorrhages, decreased electroretinogram responses, ultrastructural damage to the retina, cataract and epiretinal membrane formation following the intravitreal administration of dispase [39–41]. Therefore, these more recent studies bring into question the safety of this approach.

Hyaluronidase

Gottlieb et al. [42] investigated the effects of intravitreal hyaluronidase on the rabbit eye and concluded that it induced partial vitreolysis. A subsequent study by Harooni et al. [43] showed that hyaluronidase induced PVD, but this finding was not substantiated by other investigators [44]. A possible explanation for these differences is that preparations of hyaluronidase are often contaminated with significant amounts of proteolytic enzymes and that it was non- specific proteolysis that induced PVD in the Harooni study. A highly purified ovine hyaluronidase Vitrase™ (ISTA Pharmaceuticals, Calif., USA) has been developed for clinical use. Two phase III randomised controlled trials have shown that a single intravitreal injection of Vitrase is safe and demonstrated efficacy in aiding the clearance of vitreous haemorrhage [45, 46]. These studies did not report whether Vitrase induced PVD, but there was no propensity for retinal detachment in the Vitrase-treated group. A variable response was observed to Vitrase so a predictive model has been developed to identify those subjects that are most likely to have vitreous haemorrhage clearance by three months by analysing the patients at 1 month following treatment [47].

Chondroitinase

Intravitreal injection of chondroitinase (chondroitin ABC lyase) was reported to induce vitreoretinal separation in primates without any observed toxicity [48]. Following to this, a phase I clinical trial was initiated but the results were never published. Subsequently, the effects of chondroitin ABC lyase were investigated in porcine eyes and were reported to facilitate vitrectomy [49], but in a further study this enzyme was not observed to facilitate vitreoretinal separation during vitrectomy [50].

References

1 Bishop PN: Structural macromolecules and supramolecular organisation of the vitreous gel. Prog Retin Eye Res 2000;19:323–344.

2 Balazs EA, Denlinger JL: Aging changes in the vitreus; in Dismukes K, Sekular R (eds): Aging and Human Visual Function. New York, Liss, 1982, pp 45–57.

3 Wang J, McLeod D, Henson DB, Bishop PN: Age-dependent changes in the basal retinovitreous adhesion. Invest Ophthalmol Vis Sci 2003;44: 1793–1800.

4 Ponsioen TL, van der Worp RJ, van Luyn MJ, Hooymans JM, Los LI: Packages of vitreous collagen (type II) in the human retina: an indication of postnatal collagen turnover? Exp Eye Res 2005;80: 643–650.

5 Itakura H, Kishi S, Kotajima N, Murakami M: Vitreous collagen metabolism before and after vitrectomy. Graefes Arch Clin Exp Ophthalmol 2005; 243:994–998.

6 Bos KJ, Holmes DF, Kadler KE, McLeod D, Morris NP, Bishop PN: Axial structure of the heterotypic collagen fibrils of vitreous humour and cartilage. J Mol Biol 2001;306:1011–1022.

7 Reardon AJ, Le Goff M, Briggs MD, McLeod D, Sheehan JK, Thornton DJ, Bishop PN: Identification in vitreous and molecular cloning of opticin, a novel member of the family of leucine-rich repeat proteins of the extracellular matrix. J Biol Chem 2000; 275:2123–2129.

8 Ramesh S, Bonshek RE, Bishop PN: Immunolocalisation of opticin in the human eye. Br J Ophthalmol 2004;88:697–702.

9 Le Goff MM, Hindson VJ, Jowitt TA, Scott PG, Bishop PN: Characterization of opticin and evidence of stable dimerization in solution. J Biol Chem 2003;278:45280–45287.

10 Hindson VJ, Gallagher JT, Halfter W, Bishop PN: Opticin binds to heparan and chondroitin sulfate proteoglycans. Invest Ophthalmol Vis Sci 2005;46: 4417–4423.

11 Reardon A, Heinegard D, McLeod D, Sheehan JK, Bishop PN: The large chondroitin sulphate proteoglycan versican in mammalian vitreous. Matrix Biol 1998;17:325–333.

12 Bos KJ, Holmes DF, Meadows RS, Kadler KE, McLeod D, Bishop PN: Collagen fibril organisation in mammalian vitreous by freeze etch/rotary shadowing electron microscopy. Micron 2001;32:301–306.

13 Bishop PN, Holmes DF, Kadler KE, McLeod D, Bos KJ: Age-related changes on the surface of vitreous collagen fibrils. Invest Ophthalmol Vis Sci 2004;45: 1041–1046.

14 Bishop PN, McLeod D, Reardon A: Effects of hyaluronan lyase, hyaluronidase, and chondroitin ABC lyase on mammalian vitreous gel. Invest Ophthalmol Vis Sci 1999;40:2173–2178.

15 Spencer LM, Foos RY: Paravascular vitreoretinal attachments: role in retinal tears. Arch Ophthalmol 1970;84:557–564.

16 Fukai N, Eklund L, Marneros AG, et al: Lack of collagen XVIII/endostatin results in eye abnormalities. EMBO J 2002;21:1535–1544.

17 Sebag J, Balazs EA: Morphology and ultrastructure of human vitreous fibers. Invest Ophthalmol Vis Sci 1989;30:1867–1871.

18 Los LI, van der Worp RJ, van Luyn MJ, Hooymans JM: Age-related liquefaction of the human vitreous body: LM and TEM evaluation of the role of proteoglycans and collagen. Invest Ophthalmol Vis Sci 2003;44:2828–2833.

19 Goes RM, Nader HB, Porcionatto MA, Haddad A, Laicine EM: Chondroitin sulfate proteoglycans are structural renewable constituents of the rabbit vitreous body. Curr Eye Res 2005;30:405–413.

20 Sebag J: Anomalous posterior vitreous detachment: a unifying concept in vitreo-retinal disease. Graefes Arch Clin Exp Ophthalmol 2004;242:690–698.

21 Uemura A, Nakamura M, Kachi S, Nishizawa Y, Asami T, Miyake Y, Terasaki H: Effect of plasmin on laminin and fibronectin during plasmin-assisted vitrectomy. Arch Ophthalmol 2005;123:209–213.

22 Takano A, Hirata A, Inomata Y, Kawaji T, Nakagawa K, Nagata S, Tanihara H: Intravitreal plasmin injection activates endogenous matrix metalloproteinase-2 in rabbit and human vitreous. Am J Ophthalmol 2005;140:654–660.

23 Brown DJ, Bishop P, Hamdi H, Kenney MC: Cleavage of structural components of mammalian vitreous by endogenous matrix metalloproteinase-2. Curr Eye Res 1996;15:439–445.

24 Gandorfer A, Priglinger S, Schebitz K, Hoops J, Ulbig M, Ruckhofer J, Grabner G, Kampik A: Vitreoretinal morphology of plasmin-treated human eyes. Am J Ophthalmol 2002;133:156–159.

25 Sebag J, Ansari RR, Suh KI: Pharmacologic vitreolysis with microplasmin increases vitreous diffusion coefficients. Graefes Arch Clin Exp Ophthalmol 2007;245:576–580.

26 Verstraeten TC, Chapman C, Hartzer M, Winkler BS, Trese MT, Williams GA: Pharmacologic induction of posterior vitreous detachment in the rabbit. Arch Ophthalmol 1993;111:849–854.

27 Margherio AR, Margherio RR, Hartzer M, Trese MT, Williams GA, Ferrone PJ: Plasmin enzyme-assisted vitrectomy in traumatic pediatric macular holes. Ophthalmology 1998;105:1617–1620.

28 Trese MT, Williams GA, Hartzer MK: A new approach to stage 3 macular holes. Ophthalmology 2000;107:1607–1611.

29 Williams JG, Trese MT, Williams GA, Hartzer MK: Autologous plasmin enzyme in the surgical management of diabetic retinopathy. Ophthalmology 2001;108:1902–1905; discussion 1905–1906.

30 Sakuma T, Tanaka M, Inoue J, Mizota A, Souri M, Ichinose A: Use of autologous plasmin during vitrectomy for diabetic maculopathy. Eur J Ophthalmol 2006;16:138–140.

31 Rizzo S, Pellegrini G, Benocci F, Belting C, Baicchi U, Vispi M: Autologous plasmin for pharmacologic vitreolysis prepared 1 hour before surgery. Retina 2006;26:792–796.

32 Gandorfer A, Rohleder M, Sethi C, Eckle D, Welge-Lussen U, Kampik A, Luthert P, Charteris D: Posterior vitreous detachment induced by microplasmin. Invest Ophthalmol Vis Sci 2004;45:641–647.

33 Sakuma T, Tanaka M, Mizota A, Inoue J, Pakola S: Safety of in vivo pharmacologic vitreolysis with recombinant microplasmin in rabbit eyes. Invest Ophthalmol Vis Sci 2005;46:3295–3299.

34 Hesse L, Chofflet J, Kroll P: Tissue plasminogen activator as a biochemical adjuvant in vitrectomy for proliferative diabetic vitreoretinopathy. Ger J Ophthalmol 1995;4:323–327.

35 Le Mer Y, Korobelnik JF, Morel C, Ullern M, Berrod JP: TPA-assisted vitrectomy for proliferative diabetic retinopathy: results of a double-masked, multicenter trial. Retina 1999;19:378–382.

36 Stenn KS, Link R, Moellmann G, Madri J, Kuklinska E: Dispase, a neutral protease from Bacillus polymyxa, is a powerful fibronectinase and type IV collagenase. J Invest Dermatol 1989;93:287–290.

37 Tezel TH, Del Priore LV, Kaplan HJ: Posterior vitreous detachment with dispase. Retina 1998;18:7–15.

38 Oliveira LB, Tatebayashi M, Mahmoud TH, Blackmon SM, Wong F, McCuen BW 2nd: Dispase facilitates posterior vitreous detachment during vitrectomy in young pigs. Retina 2001;21:324–331.

39 Jorge R, Oyamaguchi EK, Cardillo JA, Gobbi A, Laicine EM, Haddad A: Intravitreal injection of dispase causes retinal hemorrhages in rabbit and human eyes. Curr Eye Res 2003;26:107–112.

40 Wang F, Wang Z, Sun X, Xu X, Zhang X: Safety and efficacy of dispase and plasmin in pharmacologic vitreolysis. Invest Ophthalmol Vis Sci 2004;45:3286–3290.

41 Zhu D, Chen H, Xu X: Effects of intravitreal dispase on vitreoretinal interface in rabbits. Curr Eye Res 2006;31:935–946.

42 Gottlieb JL, Antoszyk AN, Hatchell DL, Saloupis P: The safety of intravitreal hyaluronidase: a clinical and histologic study. Invest Ophthalmol Vis Sci 1990;31:2345–2352.

43 Harooni M, McMillan T, Refojo M: Efficacy and safety of enzymatic posterior vitreous detachment by intravitreal injection of hyaluronidase. Retina 1998;18:16–22.

44 Hikichi T, Kado M, Yoshida A: Intravitreal injection of hyaluronidase cannot induce posterior vitreous detachment in the rabbit. Retina 2000;20:195–198.

45 Kuppermann BD, Thomas EL, de Smet MD, Grillone LR: Safety results of two phase III trials of an intravitreous injection of highly purified ovine hyaluronidase (Vitrase) for the management of vitreous hemorrhage. Am J Ophthalmol 2005;140:585–597.

46 Kuppermann BD, Thomas EL, de Smet MD, Grillone LR: Pooled efficacy results from two multinational randomized controlled clinical trials of a single intravitreous injection of highly purified ovine hyaluronidase (Vitrase) for the management of vitreous hemorrhage. Am J Ophthalmol 2005;140:573–584.

47 Bhavsar A, Grillone L, McNamara T, Gow J, Hochberg A, Pearson R: Predicting response of vitreous hemorrhage and outcome of photocoagulation after intravitreous injection of highly purified ovine hyaluronidase in patients with diabetes. Invest Ophthalmol Vis Sci 2008;49:4219–4225.

48 Hageman GS, Russell SR: Chondroitinase-mediated disinsertion of the primate vitreous body. Invest Ophthalmol Vis Sci 1994;35:1260.

49 Staubach F, Nober V, Janknecht P: Enzyme-assisted vitrectomy in enucleated pig eyes: a comparison of hyaluronidase, chondroitinase, and plasmin. Curr Eye Res 2004;29:261–268.

50 Hermel M, Schrage NF: Efficacy of plasmin enzymes and chondroitinase ABC in creating posterior vitreous separation in the pig: a masked, placebo-controlled in vivo study. Graefes Arch Clin Exp Ophthalmol 2007;245:399–406.

Paul N. Bishop
Manchester Royal Eye Hospital
Oxford Road
Manchester M13 9WH (UK)
Tel. +44 161 275 5755, Fax +44 161 275 3938, E-Mail paul.bishop@manchester.ac.uk

Gandorfer A (ed): Pharmacology and Vitreoretinal Surgery.
Dev Ophthalmol. Basel, Karger, 2009, vol 44, pp 20–25

Hyaluronidase for Pharmacologic Vitreolysis

Raja Narayanan[a] · Baruch D. Kuppermann[b]

[a]Smt Kanuri Santhamma Vitreoretina Center, L.V. Prasad Eye Institute, Hyderabad, India; [b]UC Irvine Medical Center, Ophthalmology Clinic, Orange, Calif., USA

Abstract

Various pharmacologic vitreolysis agents, including hyaluronidase, urea, plasmin, dispase, tissue plasminogen activator and chondroitinase have been tested. Pharmacologic vitreolysis can avoid the complications of surgery such as cataract, endophthalmitis, retinal hemorrhage, tear or detachment, and anesthesia related complications. Hyaluronan is a major macromolecule of vitreous. It is a long, unbranched polymer of repeating disaccharide (glucuronic acid β (1,3)-N-acetylglucosamine) moieties linked by β 1–4 bonds. Hyaluronan is covalently linked to a protein core, to form a proteoglycan. It plays a pivotal role in stabilizing the vitreous gel. Hyaluronidase cleaves glycosidic bonds of hyaluronic acid and, to a variable degree, other acid mucopolysaccharides of the connective tissue. Dissolution of the hyaluronic acid and collagen complex results in decreased viscosity of the extracellular matrix. This in turn increases the diffusion rate of erythrocytes and exudates along with phagocytes through the vitreous and facilitates red blood cell lysis and phagocytosis.

Vitreous hemorrhage is a major cause of vision loss. The most common cause of vitreous hemorrhage is diabetic retinopathy, which can also lead to tractional retinal detachment [1]. Enzymatic vitreolysis offers several advantages over conventional surgery, including the ability to treat the hemorrhage earlier, in an office setting, and with lower costs. Pharmacologic vitreolysis can also avoid the complications of surgery such as cataract, endophthalmitis, retinal hemorrhage, tear or detachment, and anesthesia related complications. A number of vitreolytic enzymes have been investigated, including hyaluronidase, plasmin dispase, tissue plasminogen activator, and chondroitinase, with varying effects. Such pharmacologic therapies rely upon a good understanding of the biochemical composition and organization of vitreous.

Ultrastructural studies have shown that the adult human vitreous contains fine, parallel fibers of collagen coursing in an anteroposterior direction [2, 3]. Although several agents have been investigated in animal models, only a few have been employed in humans. Chondroitinase and hyaluronidase have been tested in clinical trials, but the former never entered phase II testing. Hyaluronidase (Vitrase) failed a phase III US Food and Drug Administration (FDA) trial conducted in the United States. The Vitrase clinical trial results will be discussed in more detail later in this chapter.

Hyaluronan

Hyaluronan is a major macromolecule of vitreous [3]. Hyaluronan is a long, unbranched polymer of repeating disaccharide (glucuronic acid β (1,3)-N-acetylglucosamine) moieties linked by β 1–4 bonds. It is a linear, left-handed, threefold helix with a rise per disaccharide on the helix axis of 0.98 nm. The sodium salt of hyaluronan has a molecular weight of 3 to 4.5×10^6 in normal human vitreous. Hyaluronan is not normally a free polymer in vivo, but it is covalently linked to a protein core, to form a proteoglycan. Hyaluronan plays a pivotal role in stabilizing the vitreous gel.

Collagen

Studies have shown that vitreous contains collagen type II, a hybrid of types V/XI, and type IX collagen in a molar ratio of 75:10:15, respectively [4]. Vitreous collagens are organized into fibrils with type V/XI residing in the core, type II collagen surrounding the core, and type IX collagen on the surface of the fibril. The fibrils are 7–28 nm in diameter, but their length in situ is unknown.

Supramolecular Organization

Vitreous is a dilute meshwork of collagen fibrils interspersed with extensive arrays of hyaluronan molecules. The collagen fibrils provide a scaffold-like structure that is 'inflated' by the hydrophilic hyaluronan. If collagen is removed, the remaining hyaluronan forms a viscous solution. However, if hyaluronan is removed, the gel shrinks but is not destroyed. Electrostatic binding occurs between the negatively charged hyaluronan and the positively charged collagen in vitreous [5].

Studies have shown that the chondroitin sulfate chains of type IX collagen bridge between adjacent collagen fibrils in a ladder-like configuration spacing them apart [6]. Such spacing is necessary for vitreous transparency, because keeping vitreous collagen fibrils separated by at least one wavelength of incident light minimizes light scattering, allowing the unhindered transmission of light to the retina for photoreception. Bishop [4] proposed that the leucine-rich repeat protein opticin is the predominant structural protein responsible for short-range spacing of collagen fibrils.

Numerous changes occur in the structure of vitreous with age. There is a significant decrease in the gel volume and an increase in the liquid volume of human vitreous.

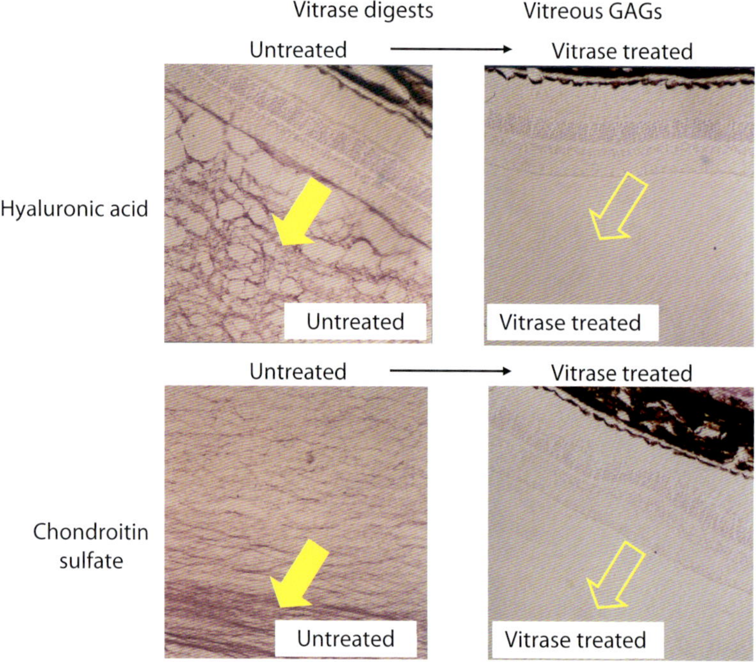

Fig. 1. Studies in rabbit vitreous injected with Vitrase show digestion of hyaluronic acid and chondroitin sulfate compared to untreated controls.

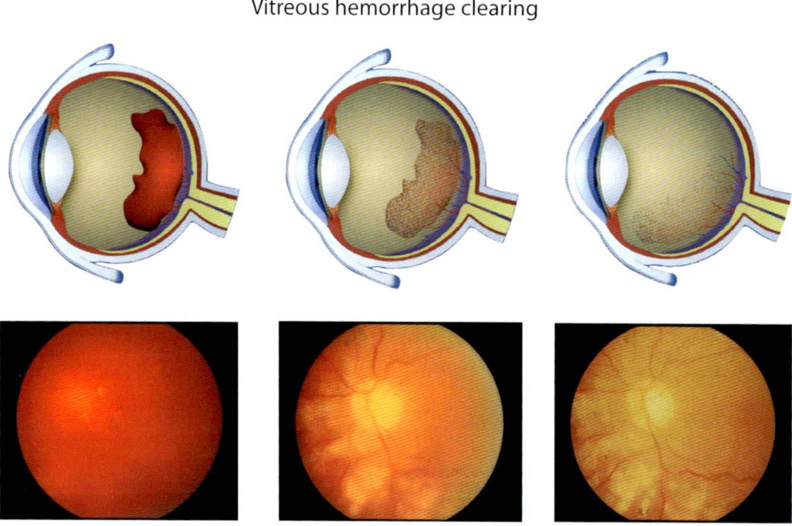

Fig. 2. Clearing of dense vitreous hemorrhage of greater than one month duration in a patient with proliferative diabetic retinopathy. Schematic sequence above and photographic sequence below showing degree of vitreous hemorrhage immediately prior to injection of 55IU Vitrase, then one week post injection showing partial clearing, and finally two weeks post injection with near complete clearance of vitreous hemorrhage.

Derangement of the normal hyaluronan/collagen association results in the simultaneous formation of liquid vitreous and aggregation of collagen fibrils into bundles of parallel fibrils seen macroscopically as large fibers [7]. In the posterior vitreous, such age-related changes form large pockets of liquid vitreous.

Vitreoretinal adhesion at the posterior pole is more fascial than focal (at the disc, fovea, and along retinal blood vessels), consistent with the concept that the two tissues are held together by an extracellular matrix tissue [8]. With age, there is weakening of vitreoretinal adhesion, most likely due to biochemical alterations in the extracellular matrix tissue at the vitreoretinal interface. Studies using lectin probes have identified that one component of the extracellular matrix at the vitreoretinal interface [galactose-β (1,3)-N-acetylglucosamine] is present in youth but absent in adults [9]. This difference and others may play a role in the observed weakening of the vitreoretinal interface during aging.

Anomalous PVD

Anomalous PVD results when the extent of vitreous liquefaction exceeds the degree of vitreoretinal interface weakening, resulting in traction exerted at the vitreoretinal interface [10]. There can be various untoward effects of anomalous PVD. Effects upon the retina include hemorrhage, retinal tears and detachment, vitreomacular traction syndromes, and some cases of diffuse diabetic macular edema. Proliferative diabetic retinopathy can be greatly aggravated by anomalous PVD. Effects upon vitreous involve posterior vitreoschisis, where splitting of the posterior vitreous cortex and forward displacement of the vitreous body leave the outer layer of the split posterior vitreous cortex still attached to the retina. This can result in macular pucker, contribute to macular holes, or complicate proliferative diabetic retinopathy.

Pharmacologic Vitreolysis

Various pharmacologic vitreolysis agents, including hyaluronidase, urea, plasmin, dispase, tissue plasminogen activator and chondroitinase have been tested to date [11], but little is known about the exact mechanism of action of these substances, and so far none has met with sufficient success to result in widespread use.

Hyaluronidase

Hyaluronidase cleaves glycosidic bonds of hyaluronic acid and, to a variable degree, other acid mucopolysaccharides of the connective tissue. Dissolution of the hyaluronic acid and collagen complex results in decreased viscosity of the extracellular matrix. This in turn increases the diffusion rate of erythrocytes and exudates along with phagocytes through the vitreous and facilitates red blood cell lysis and phagocytosis.

Vitrase (Hyaluronidase)
Vitrase was studied in a large, multicenter, randomized controlled trial [12]. A total of 1,125 patients with persistent vitreous hemorrhage were randomized to 55 IU (n = 365), 75 IU (n = 377), and saline (n = 383). The percentage of patients reaching primary efficacy by month 1 in the 55 IU, 75 IU, and saline groups was 13.2, 10.6, and 5.5% (p = 0.001 and p = 0.010, respectively). By month 3, 32.9, 30.5, and 25.6% of patients treated with 55 IU, 75 IU, and saline reached the primary efficacy endpoint (p = 0.025, p = 0.144, respectively). A statistically significant improvement in BCVA was seen. By month 3, 44.9 and 43.5% of patients in the 55 and 75 IU treatment groups had at least a 3-line improvement in BCVA compared with 34.5% of patients in the saline group (p = 0.004 and p = 0.011, respectively). The analysis of the investigator graded reduction in vitreous hemorrhage density also showed statistical significance in the 55 and 75 IU dose groups compared with saline by months 1, 2 and 3. In the analysis of the clinical assessment of therapeutic utility (clearance of the hemorrhage sufficient to diagnose the underlying pathology), there was a significant difference for patients in the 55 and 75 IU treatment groups compared with saline-treated patients by months 1, 2 and 3. By month 3, 40.8 and 39.3% of patients in the 55 IU (p = 0.001) and 75 IU (p = 0.002) groups, respectively, reached this endpoint compared with 28.5% of patients in the saline group.

In summary, data from the phase III clinical trials demonstrated that intravitreous injection of ovine hyaluronidase was efficacious and had a favorable safety profile. The primary ocular adverse event associated with the use of ovine hyaluronidase was acute, self-limited anterior uveitis. This iritis resolved with or without treatment, and without sequelae. Additionally, clinical experience has demonstrated that these injections do not prohibit or complicate subsequent vitrectomies when they are required.

References

1 Spraul CW, Grossniklaus HE: Vitreous hemorrhage. Surv Ophthalmol 1997;42:3–39.
2 Sebag J, Balazs EA: Pathogenesis of CME: anatomic considerations of vitreo-retinal adhesions. Surv Opthalmol 1984;28:493–498.
3 Sebag J: Molecular biology of pharmacologic vitreolysis. Trans Am Ophthalmol Soc 2005;103:473–494.
4 Bishop PN: Structural macromolecules and supramolecular organization of the vitreous gel. Prog Ret Eye Res 2000;19:323–344.
5 Comper WD, Laurent TC: Physiological functions of connective tissue polysaccharides. Physiol Rev 1978;58:255–315.
6 Scott JE, Chen Y, Brass A: Secondary and tertiary structures involving chondroitin and chondroitin sulphate in solution, investigated by rotary shadowing electron microscopy and computer simulation. Eur J Biochem 1992;209:675–680.
7 Sebag J, Balazs EA: Morphology and ultrastructure of human vitreous fibers. Invest Ophthalmol Vis Sci 1989;30:1867–1871.

8 Sebag J: Age-related differences at the human vitreo-retinal interface. Arch Ophthalmol 1991;109:966–971.

9 Russell SR, Shepherd JD, Hageman GS: Distribution of glycoconjugates in the human internal limiting membrane. Invest Ophthalmol Vis Sci 1991;32:1986–1995.

10 Sebag J. Anomalous PVD: a unifying concept in vitreo-retinal disease. Graefe's Arch Clin Exp Ophthalmol 2004;242:690–698.

11 Bhisitkul RB: Anticipation for enzymatic vitreolysis. Br J Ophthalmol 2001;85:1–2.

12 Kuppermann BD, Thomas EL, de Smet M, Grillone LR: Pooled efficacy results from two multinational randomized controlled trials of a single intravitreous injection of highly purified ovine hyaluronidase for the management of vitreous hemorrhage. Am J Ophthalmol 2005;140:573–584.

Raja Narayanan, MD
Smt Kanuri Santhamma Vitreoretina Center, L.V. Prasad Eye Institute
L.V. Prasad Marg, Banjara Hills
Hyderabad 500034 (India)
E-Mail drnons@yahoo.com

Gandorfer A (ed): Pharmacology and Vitreoretinal Surgery.
Dev Ophthalmol. Basel, Karger, 2009, vol 44, pp 26–30

Microplasmin-Assisted Vitrectomy

Arnd Gandorfer

Vitreoretinal and Pathology Unit, Augenklinik der Ludwig-Maximilians-Universität, Munich, Germany

Abstract

Microplasmin (Thrombogenics Ltd., Leuven, Belgium) is a recombinant molecule consisting of the catalytic domain of human plasmin. It is highly characterized, and supplied in a stabilized form, simplifying storage and administration. Microplasmin shares the same catalytic properties like human plasmin, but it is much more stable compared to plasmin. It has been shown previously that both plasmin and microplasmin are capable of inducing PVD. This chapter focuses on one of the most promising current concepts of pharmacologic vitreolysis, i.e. microplasmin-assisted vitrectomy. We report on the preclinical work of plasmin and microplasmin which lead to the clinical investigation of microplasmin in different clinical trials, the so-called MIVI trials.

Plasmin, a non-specific serine protease mediating the fibrinolytic process, acts also on a variety of glycoproteins including laminin and fibronectin. Both of them are present at the vitreoretinal interface, and they are believed to chemically bind the vitreous to the internal limiting membrane (ILM) of the retina [1–3].

In 1993, PVD could be achieved in rabbit eyes by intravitreal injection of the enzyme followed by vitrectomy [4]. In 1999, Hikichi et al. [5] confirmed complete PVD after injection of 1U plasmin and 0.5 ml SF_6 gas in the rabbit model, without evidence of retinal toxicity.

Unfortunately, plasmin is not yet available for clinical application, and alternative strategies have been pursued to administer the enzyme in vitreoretinal surgery. Tissue plasminogen activator was injected into the vitreous in an attempt to generate plasmin by intravitreal activation of endogenous plasminogen. In an animal model in rabbit eyes, complete PVD was observed in all eyes treated with 25 μg tissue plasminogen activator [6]. Breakdown of the blood-retinal barrier was necessary to allow plasminogen to enter the vitreous, and this was induced by cryocoagulation [6]. In

The author is co-founder of the Microplasmin Study Group and has a financial interest in pharmacologic vitreolysis.

two clinical pilot studies, 25 µg tissue plasminogen activator was injected into the vitreous of patients with proliferative diabetic retinopathy 15 min before vitrectomy [7, 8]. The results of both studies, however, were contradictory in terms of PVD induction and clinical benefit. Recently, Peyman's group [9] demonstrated PVD induction in rabbit eyes by an intravitreal administration of recombinant lysine-plasminogen and recombinant urokinase.

Autologous plasminogen purified from the patient's own plasma by affinity chromatography was converted to plasmin by streptokinase in vitro. 0.4 U of autologous plasmin enzyme (APE) was injected into the vitreous in patients with pediatric macular holes, diabetic retinopathy, and stage 3 idiopathic macular holes, followed by vitrectomy after 15 min [10–12]. All APE-treated eyes achieved spontaneous or easy removal of the posterior hyaloid including one eye that had vitreoschisis over areas of detached retina.

Microplasmin (Thrombogenics Ltd., Leuven, Belgium) is a recombinant product which contains the catalytic domain of human plasmin only [13]. It shares the catalytic properties of human plasmin, but is much more stable than the original molecule. Microplasmin is highly characterized. It is supplied in powder form which simplifies storage and administration.

As discussed in the first chapter, the goal of enzymatic vitreolysis is to manipulate the vitreous collagen pharmacologically, both centrally achieving liquefaction (synchisis), as well as along the vitreoretinal interface to induce posterior vitreous detachment (PVD, syneresis), and to create a cleavage plane more safely and cleaner than can currently be achieved by mechanical means [14–17].

Preclinical Evaluation of Plasmin-Assisted Vitrectomy

We investigated the effect of plasmin in porcine postmortem eyes and in human donor eyes. In porcine eyes, we observed a dose-dependent separation of the vitreous cortex from the ILM after intravitreal injection, without additional vitrectomy or gas injection [18]. In scanning electron microscopy, a bare ILM was achieved with 1 U of porcine plasmin 60 min after injection, and with 2 U of plasmin 30 and 60 min after injection, respectively. In control fellow eyes which were injected with BSS, the cortical vitreous remained attached to the retina [18]. For these experiments, we have developed a straightforward grading system that allows for easy and reliable quantification of cortical vitreous remnants at the ILM when scanning electron microscopy is performed [18].

In human donor eyes, 2 U of human plasmin from pooled plasma achieved complete PVD 30 min after injection, whereas the vitreoretinal surface of the fellow eyes was covered by collagen fibrils [19]. In both studies, transmission electron microscopy revealed a clean and perfectly preserved ILM in plasmin-treated eyes, and no evidence of inner retinal damage was seen [18, 19]. Li et al. [20] confirmed these results, and reported a reduced immunoreactivity of the vitreoretinal interface for laminin and fibronectin following plasmin application.

In an experimental setting simulating the application of plasmin as an adjunct to vitrectomy, we injected human donor eyes with 1 U of plasmin, followed by vitrectomy 30 min thereafter [21]. All plasmin-treated eyes showed complete PVD, whereas the control eyes which were vitrectomized conventionally had various amounts of the cortical vitreous still present at the vitreoretinal interface [21].

Preclinical Evaluation of Microplasmin-Assisted Vitrectomy

In 2001, we approached Thrombogenics, a drug development company which had manufactured recombinant microplasmin for clinical investigation in patients with stroke and peripheral artery occlusive disease. Given our preclinical work with human plasmin from pooled plasma, we tested microplasmin in human postmortem eyes in the same manner as we did before with human plasmin. There was a clear dose-response relationship of microplasmin comparable to that of human plasmin. Complete vitreoretinal separation was possible without affecting retinal morphology [22]. We went on testing the substance in the cat model in vivo (fig. 1). No alteration in retinal ultrastructure was seen, and there was no change in the antigenicity of neurons and glial cells [22]. This was important to know as Mueller cells are very sensitive to any form of ocular trauma and intraocular surgery. Further animal studies conducted by the manufacturer followed in different species. Formal toxicology testing was performed, and microplasmin entered the clinical phase.

Clinical Evaluation of Microplasmin-Assisted Vitrectomy

We designed the first clinical study investigating the effect of microplasmin in combination with vitrectomy (MIVI-I). This phase II study aimed at detemining the safety of microplasmin following intravitreal injection. 60 patients were included, and all received an intravitreal injection of microplasmin before vitrectomy. In the six cohorts, either the dose or the incubation time between injection and surgery were increased. In this setting, microplasmin appeared to be safe, and preliminary efficacy results were promising.

There are several other European studies underway to investigate the effect of microplasmin on vitreolysis; a nonvitrectomy trial in patients with diabetic macular edema (MIVI-II), and a combined injection/vitrectomy trial in patients with traction maculopathy (MIVI-IIt). The results of the MIVI-I study have already served as a basis for the FDA opening study (MIVI-III) which was performed in the USA. According to a press release by Thrombogenics from June 2008, approximately 30% of patients showed relief of vitreoretinal traction without the need for vitrectomy.

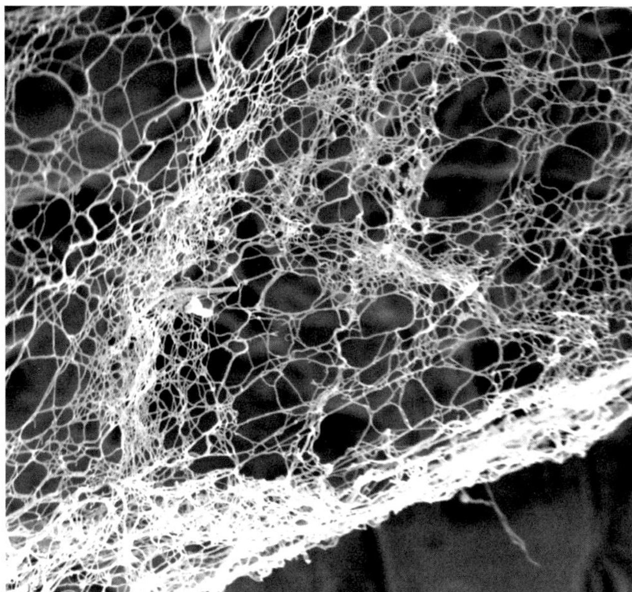

Fig. 1. Separation of the cortical vitreous from the retina in a cat eye. The vitreous cortex is seen in focus, the retina (defocused) is devoid of collagen remnants. Scanning electron microscopy. ×7,000.

Conclusion

Plasmin and microplasmin hold the promise of inducing complete PVD without causing morphologic alteration of the retina. Several independent studies confirmed a dose-dependent and complete vitreoretinal separation, associated with perfect preservation of the ultrastructure of the ILM and the retina [4, 5, 18, 19, 21, 22]. In addition, a dose-dependent liquefaction of the vitreous induced by microplasmin was demonstrated by dynamic light scattering in dissected porcine vitreous and in intact pig eyes, making plasmin and microplasmin the most promising agents for pharmacologic vitreolysis at the moment [23]. Data from the first clinical trial of microplasmin are available and are promising in terms of safety and efficacy. Randomized studies are now performed to further assess the safety and efficacy of microplasmin when used as an adjunct to vitrectomy, or even as its replacement.

References

1 Kohno T, Sorgente N, Ishibashi T, Goodnight R, Ryan SJ: Immunofluorescent studies of fibronectin and laminin in the human eye. Invest Ophthalmol Vis Sci 1987;28:506–514.

2 Kohno T, Sorgente N, Goodnight R, Ryan SJ: Alterations in the distribution of fibronectin and laminin in the diabetic human eye. Invest Ophthalmol Vis Sci 1987;28:515–521.

3 Liotta LA, Goldfarb RH, Brundage R, Siegal GP, Terranova V, Garbisa S: Effect of plasminogen activator (urokinase), plasmin, and thrombin on glycoprotein and collagenous components of basement membrane. Cancer Res 1981;41:4629–4636.

4 Verstraeten TC, Chapman C, Hartzer M, Winkler BS, Trese MT, Williams GA: Pharmacologic induction of posterior vitreous detachment in the rabbit. Arch Ophthalmol 1993;111:849–854.
5 Hikichi T, Yanagiya N, Kado M, Akiba J, Yoshida A: Posterior vitreous detachment induced by injection of plasmin and sulfur hexafluoride in the rabbit vitreous. Retina 1999;19:55–58.
6 Hesse L, Nebeling B, Schroeder B, Heller G, Kroll P: Induction of posterior vitreous detachment in rabbits by intravitreal injection of tissue plasminogen activator following cryopexy. Exp Eye Res 2000;70: 31–39.
7 Hesse L, Chofflet J, Kroll P: Tissue plasminogen activator as a biochemical adjuvant in vitrectomy for proliferative diabetic vitreoretinopathy. Ger J Ophthalmol 1995;4:323–327.
8 Le Mer Y, Korobelnik JF, Morel C, Ullern M, Berrod JP: TPA-assisted vitrectomy for proliferative diabetic retinopathy: results of a double-masked, multicenter trial. Retina 1999;19:378–382.
9 Men G, Peyman GA, Genaidy M, Kuo PC, Ghahramani F, Blake DA, Bezerra Y, Naaman G, Figueiredo E: The role of recombinant lysine-plasminogen and recombinant urokinase and sulfur hexafluoride combination in inducing posterior vitreous detachment. Retina 2004;24:199–209.
10 Margherio AR, Margherio RR, Hartzer M, Trese MT, Williams GA, Ferrone PJ: Plasmin enzyme-assisted vitrectomy in traumatic pediatric macular holes. Ophthalmology 1998;105:1617–1620.
11 Trese MT, Williams GA, Hartzer MK: A new approach to stage 3 macular holes. Ophthalmology 2000;107:1607–1611.
12 Williams JG, Trese MT, Williams GA, Hartzer MK: Autologous plasmin enzyme in the surgical management of diabetic retinopathy. Ophthalmology 2001;108:1902–1905.
13 Nagai N, Demarsin E, Van Hoef B, Wouters S, Cingolani D, Laroche Y, Collen D: Recombinant human microplasmin: production and potential therapeutic properties. J Thromb Haemost 2003;1: 307–313.
14 Trese M: Enzymatic-assisted vitrectomy. Eye 2002; 16:365–368.
15 Bhisitkul RB: Anticipation for enzymatic vitreolysis. Br J Ophthalmol 2001;85:1–2.
16 Sebag J: Pharmacologic vitreolysis. Retina 1998;18: 1–3.
17 Sebag J: Is pharmacologic vitreolysis brewing? Retina 2002;22:1–3.
18 Gandorfer A, Putz E, Welge-Lussen U, Gruterich M, Ulbig M, Kampik A: Ultrastructure of the vitreoretinal interface following plasmin assisted vitrectomy. Br J Ophthalmol 2001;85:6–10.
19 Gandorfer A, Priglinger S, Schebitz K, Hoops J, Ulbig M, Ruckhofer J, Grabner G, Kampik A: Vitreoretinal morphology of plasmin-treated human eyes. Am J Ophthalmol 2002;133:156–159.
20 Li X, Shi X, Fan J: Posterior vitreous detachment with plasmin in the isolated human eye. Graefes Arch Clin Exp Ophthalmol 2002;240:56–62.
21 Gandorfer A, Ulbig M, Kampik A: Plasmin-assisted vitrectomy eliminates cortical vitreous remnants. Eye 2002;16:95–97.
22 Gandorfer A, Rohleder M, Sethi C, Eckle D, Welge-Lussen U, Kampik A, Luthert P, Charteris D: Posterior vitreous detachment induced by microplasmin. Invest Ophthalmol Vis Sci 2004;45:641–647.
23 Sebag J, Ansari RR, Suh KI: Pharmacologic vitreolysis with microplasmin increases vitreous diffusion coefficients. Graefes Arch Clin Exp Ophthalmol 2007;245:576–580.

Arnd Gandorfer, MD, PD, FEBO
Vitreoretinal and Pathology Unit, Augenklinik der Ludwig-Maximilians-Universität
Mathildenstrasse 8
DE–80336 Munich (Germany)
Tel. +49 089 5160 3800, Fax +49 089 5160 4778, E-Mail arnd.gandorfer@med.uni-muenchen.de

Gandorfer A (ed): Pharmacology and Vitreoretinal Surgery.
Dev Ophthalmol. Basel, Karger, 2009, vol 44, pp 31–36

Pharmacologic Vitreodynamics and Molecular Flux

David T. Goldenberg[a] · Michael T. Trese[a,b]

[a]Associated Retinal Consultants, Royal Oak, Mich., and [b]Eye Research Institute, Oakland University Rochester, Mich., USA

Abstract

Several enzymatic agents, such as autologous plasmin enzyme and recombinant microplasmin, are able to cause vitreous liquefaction and a complete posterior vitreous detachment (PVD). Advancements in research have helped to explain the complex interactions that occur in the vitreous cavity after a PVD is created. The development of a PVD is a dynamic process that is thought to have a larger impact on the vitreous cavity milieu than just a separation of the posterior cortical vitreous from the retina. Pharmacologic vitreodynamics attempts to explain the mechanical and biochemical changes that occur at the vitreoretinal junction after a PVD is formed. The flow of molecules into and out of the vitreous cavity and across the vitreoretinal junction is thought to be influenced by the presence or absence of a PVD. A microplasmin-induced PVD has been shown to alter the vitreous levels of several molecules, and a PVD may have a protective role in multiple diseases. Significant progress has been made in the field of pharmacologic vitreodynamics. As we improve our understanding of the molecular flux in the vitreous cavity, pharmacologic vitreodynamics will likely become more important as it may allow for improved manipulation of intravitreal molecules.

During the last 10 years, there has been substantial progress in our understanding of the vitreous and its role in disease pathogenesis. Manipulation of the vitreous with pharmacologic agents has improved our ability to define the complex relationships between the vitreous and the retina. Enzymatic agents such as autologous plasmin enzyme and recombinant microplasmin (ThromboGenics Ltd., Dublin, Ireland) are able to successfully cause vitreous liquefaction and a complete posterior vitreous detachment (PVD) [1–10]. Not only do these chemical agents cleave the vitreoretinal junction, but they may also help to explain some of the biochemical changes that occur in the vitreous cavity after a PVD is created. Pharmacologic vitreodynamics is a novel term that describes the use of enzymatic agents to manipulate the vitreous and,

Dr. Trese has a financial interest in plasmin and microplasmin.

more importantly, the complex mechanical and biochemical interactions at the vitreoretinal interface. Significant advancements have been made in the field of pharmacologic vitreodynamics, and this review will provide a summary of recent findings.

Molecular Flux

Drug injection studies into the vitreous cavity currently consider all vitreous cavities to be equal. In contrast, all vitreous cavities are not the same since vitreous liquefaction occurs to varying degrees in each eye. Furthermore, the posterior hyaloid may or may not be detached and the thickness and adherence of this posterior hyaloid may vary from location to location within an eye. It has been suggested that an attached posterior hyaloid (i.e. absence of a PVD) may influence the flux of molecules that cross the vitreoretinal junction [11]. It is possible that the posterior vitreous cortex creates a barrier to certain molecules in the vitreous cavity. For example, in hyaluronidase-injected cat eyes (vitreous liquefaction without a PVD) and in control eyes (no liquefaction and no PVD), the cortical vitreous appears more densely compressed at the retinal surface than in microplasmin-injected eyes which demonstrate a smooth retinal surface that is free of cortical vitreous (fig. 1) [8]. It is also possible that the connections between the vitreous and retina provide a barrier to molecular diffusion. Laminin and fibronectin are the main glycoproteins that bind vitreous collagen fibers of the posterior vitreous cortex to the internal limiting membrane (ILM) [12]. The failure of tissue plasminogen activator (tPA) to penetrate the retina has been attributed to its specific binding to fibronectin, in addition to other proteins at the vitreoretinal juncture [13]. Plasmin enzyme and microplasmin are able to hydrolyze laminin and fibronectin at the vitreoretinal junction [14, 15], and studies are currently underway to determine if a microplasmin-induced PVD increases retinal penetration of tPA. The factors that contribute to the flux of molecules into and out of the vitreous cavity are likely to be multifactorial. Nevertheless, multiple studies suggest that a PVD may play a role in this process.

PVD and Vitreous Oxygen Levels

In the nonvitrectomized eye, there is a concentration gradient of oxygen from the retina into the vitreous gel [16]. Previous reports have shown that oxygen levels in the vitreous cavity increase after pars plana vitrectomy, and this increase is sustained for several months after surgery [16, 17]. It has been suggested that the sustained increase in oxygen levels is due to the circulation of fluid in the vitreous cavity that occurs after removal of the vitreous gel [17]. Sebag et al. [9] have shown that injection of microplasmin increases the vitreous diffusion coefficient in porcine eyes using dynamic light scattering. Their results indicate increased liquefaction of the vitreous

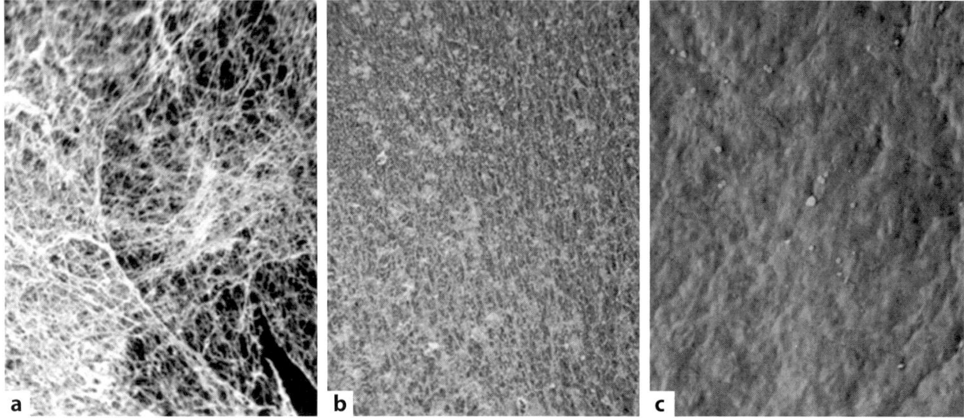

Fig. 1. Scanning electron microscopy of cat vitreoretinal interface. **a** Control showing surface with dense network of cortical vitreous. **b** After hyaluronidase injection, condensed cortical vitreous is present at the vitreoretinal interface following vitreous liquefaction without posterior vitreous separation. **c** Smooth retinal surface free of cortical vitreous following induction of a posterior vitreous detachment with microplasmin. Originally published in Retina 2007; 27: 1090-1096. Reprinted with permission from the publisher.

following microplasmin injection [9], which may allow for increased mixing and oxygen exchange within the vitreous cavity. Recently, Quiram et al. [8] used animal models to study the effect of an enzymatic-assisted PVD on the oxygen concentration in the vitreous. The authors showed that oxygen flow across the vitreoretinal interface can be altered by a microplasmin-induced PVD [8]. Specifically, they found that a microplasmin-induced PVD increases vitreous oxygen levels and the rate of oxygen exchange within the vitreous cavity [8]. This effect was not seen with hyaluronidase and liquefaction of the vitreous cavity alone [8], which suggests that oxygen diffusion may be more influenced by a PVD rather than vitreous liquefaction. Even then, the authors were unable to determine if the elevated vitreal oxygen was due to an increase in oxygen diffusion or an increase in bulk flow (mixing) of oxygen in liquefied vitreous.

In patients with proliferative diabetic retinopathy (PDR), retinal ischemia leads to elevated levels of vascular endothelial growth factor (VEGF) which induces retinal neovascularization and macular edema [18, 19]. Pan-VEGF blockade with intravitreal bevacizumab results in regression of retinal neovascularization and reduction of fluorescein angiography leakage [20]. It has been proposed that PVD formation with microplasmin could increase vitreal oxygen levels, thereby reducing the ischemic drive to produce VEGF with subsequent reduction in neovascularization [8]. Consistent with this theory is the observation that vitrectomy with complete posterior vitreous separation is believed to reduce the progression of diabetic retinopathy and macula edema in diabetic eyes [3, 21, 22]. Furthermore, a complete PVD has been found to be a strong negative risk factor for the progression of diabetic retinopathy [21].

PVD and Other Vitreous Molecules

The change in vitreous oxygen molecular flux that was demonstrated by Quiram et al. [8] may be used as a marker for the behavior of other molecules in the extracellular matrix of the vitreous cortex. Preliminary data from cat models suggest a microplasmin-induced PVD leads to decreased levels of VEGF in the vitreous [23]. In addition, we have shown that the retinal penetration of intravitreal bevacizumab is initially increased in rabbits with a microplasmin-induced PVD when compared with control eyes [24]. It is possible that a PVD may influence the concentration of many molecules and growth factors in the vitreous cavity [11]. It is unclear whether a specific relationship exists between a PVD and vitreous growth factors. Prospective clinical trials will be needed to address this theory.

Pharmacologic vitreodynamics may also have a protective role in age-related macular degeneration (AMD). A higher incidence of attached posterior vitreous has been reported to occur in patients with all types of AMD (exudative and nonexudative) [25]. Additional studies have concluded that persistent attachment of the posterior vitreous cortex to the macula may be a risk factor for the development of exudative AMD [26, 27]. Krebs et al. [26] attributed these findings to several theories: (1) chronic inflammation in the macula due to persistent vitreomacular adhesion, (2) the presence of an attached posterior vitreous prevents the normal diffusion of oxygen and other nutrients across the vitreomacular interface, and (3) the presence of an attached posterior vitreous confines VEGF and other cytokines in the macula. The association between vitreomacular adhesion and exudative AMD is probably multifactorial. However, if the diffusion of molecules across the vitroeretinal junction plays a role, then pharmacologic vitreodynamics has the potential to significantly alter the natural history of AMD.

Conclusion

The emerging field of pharmacologic vitreodynamics presents a new frontier in the management of vitreoretinal diseases. As we improve our understanding of the molecular flux in the vitreous cavity, pharmacologic vitreodynamics will likely become more important as it may allow for improved manipulation of intravitreal molecules. The use of enzymatic agents may soon enable physicians to better tailor their treatments for a variety of vitreoretinal diseases. Pharmacologic alteration of the vitreous is an evolving modality that will likely be used more frequently for therapy and preventative measures.

References

1 Trese MT, Williams GA, Hartzer MK: A new approach to stage 3 macular holes. Ophthalmology 2003;107:1607–1611.

2 Margherio AR, Margherio RR, Hartzer M, Trese MT, Williams GA, Ferrone PJ: Plasmin enzyme-assisted vitrectomy in traumatic pediatric macular holes. Ophthalmology 1998;105:1617–1620.

3 Williams JG, Trese MT, Williams GS, Hartzer MK: Autologous plasmin enzyme in the surgical management of diabetic retinopathy. Ophthalmology 2001; 108:1902–1905.

4 Asami T, Terasaki H, Kachi S, Nakamura M, Yamamura K, Nabeshima T, Miyake Y: Ultrastructure of internal limiting membrane removed during plasmin-assisted vitrectomy from eyes with diabetic macular edema. Ophthalmology 2004;111:231–237.

5 Azzolini C, D'Angelo A, Maestranzi G, Codenotti M, Della Valle P, Prati M, Brancato R: Intrasurgical plasmin enzyme in diabetic macular edema. Am J Ophthalmol 2004;138:560–566.

6 Gandorfer A, Rohleder M, Sethi C, Eckle D, Welge-Lüssen U, Kampik A, Luthert P, Charteris D: Posterior vitreous detachment induced by microplasmin. Invest Ophthalmol Vis Sci 2004;45:641–647.

7 Sakuma T, Tanaka M, Mitzota A, Inoue J, Pakola S: Safety of in vivo pharmacologic vitreolysis with recombinant microplasmin in rabbit eyes. Invest Ophthalmol Vis Sci 2005;46:3295–3299.

8 Quiram PA, Leverenz VA, Baker RM, Dang L, Giblin F, Trese M: Microplasmi-induced posterior vitreous detachment affects vitreous oxygen levels. Retina 2007;27:1090–1096.

9 Sebag J, Ansari RR, Suh KI: Pharmacologic vitreolysis with microplasminin increases vitreous diffusion coefficients. Graefe's Arch Clin Exp Ophthalmol 2007; 245:576–580.

10 Sebag J. Molecular biology of pharmacologic vitreolysis. Trans Am Ophthalmol Soc 2005;103:473–494.

11 Goldenberg DT, Trese MT: Pharmacologic vitreodynamics: what is it? Why is it important? Expert Rev Ophthalmol 2008;3:273–277.

12 Kohno T, Sorgente N, Ishibashi T, Goodnight R, Ryan S: Immunofluorescent studies of fibronectin and laminin in the human eye. Invest Ophthalmol Vis Sci 1987;28:506–514.

13 Kamei M, Misono K, Lewis H: A study of the ability of tissue plasminogen activator to diffuse into the subretinal space after intravitreal injection in rabbits. Am J Ophthalmol 1999;128:739–746.

14 Liotta LA, Goldfarb RH, Brundage R, Siegal GP, Terranova V, Garbisa S: Effect of plasminogen activator (urokinase), plasmin, and thrombin on glycoprotein and collagenous components of basement membrane. Cancer Res 1981;41:4629–4636.

15 Uemura A, Nakamura M, Kachi S, Nishizawa Y, Asami T, Miyake Y, Terasaki H: Effect of plasmin on laminin and fibronectin during plasmin-assisted vitrectomy. Arch Ophthalmol 2005;123:209–213.

16 Barbazetto IA, Liang J, Chang S, Zheng L, Spector A, Dillon JP: Oxygen tension in the rabbit lens and vitreous before and after vitrectomy. Exp Eye Res 2004;78:917–924.

17 Holekamp NM, Shui Y-B, Beebe D: Vitrectomy surgery increases oxygen exposure to the lens: A possible mechanism for nuclear cataract formation. Am J Ophthalmol 2005;139:302–310.

18 Patel JI, Tombran-Tink J, Hykin PG, Gregor ZJ, Cree IA: Vitreous and aqueous concentrations of proangiogenic, antiangiogenic factors and other cytokines in diabetic retinopathy patients with macular edema: Implications for structural differences in macular profiles. Exp Eye Res 2006;82:798–806.

19 Paques M, Massin P, Gaudric A: Growth factors and diabetic retinopathy. Diabetes Metab 1997;23:125–130.

20 Avery RL, Pearlman J, Pieramici DJ, Rabena MD, Castellarin AA, Nasir MA, Giust MJ, Wendel R, Patel A: Intravitreal bevacizumab (Avastin) in the treatment of proliferative diabetic retinopathy. Ophthalmology 2006;113:1695.e1–15.

21 Ono R, Kakehashi A, Yamagami H, Sugi N, Kinoshita N, Saito T, Tamemoto H, Kuroki M, Lshikawa SE, Kawakami M: Prospective assessment of proliferative diabetic retinopathy with observations of posterior vitreous detachment. Int Ophthalmol 2005;26:15–19.

22 Grigorian R, Bhagat N, Lanzetta P, Tutela A, Zarbin M: Pars plana vitrectomy for refractory diabetic macular edema. Semin Ophthalmol 2003;18:116–120.

23 Quiram PA, Leverenz V, Baker R, Loan D: Enzymatic induction of a posterior vitreous detachment alters molecular vitreodynamics in animal models (ARVO abstract). Invest Ophthalmol Vis Sci 2007:abstr 83.

24 Goldenberg DT, Leverenz VR, Giblin FJ, Drenser KA, Trese MT: Posterior vitreous detachment alters retinal penetration of intravitreal bevacizumab (Avastin) in rabbit eyes (ARVO abstract). Invest Ophthalmol Vis Sci 2008:abstr 5616.

25 Öndes F, Yilmaz G, Acar MA, Ünlü N, Hülya K, Arsan AK: Role of the vitreous in age-related macular degeneration. Jpn J Ophthalmol 2000;44:91–93.
26 Krebs I, Brannath W, Glittenberg C, Zeiler F, Sebag J, Binder S: Posterior vitreomacular adhesion: a potential risk factor for exudative age-related macular degeneration? Am J Ophthalmol 2007;144:741–746.
27 Quaranta-El MM, Mauget-Faÿsse M: Anomalous vitreoretinal adhesions in patients with exudative age-related macular degeneration: an OCT study. Eur J Ophthalmol 2006;16:134–137.

David T. Goldenberg, MD
Associated Retinal Consultant
3535 W. 13 Mile Rd., Suite 344
Royal Oak, MI 48073 (USA)
Tel. +1 248 288 2280, Fax +1 248 288 5644, E-Mail davidtgold@yahoo.com

Gandorfer A (ed): Pharmacology and Vitreoretinal Surgery.
Dev Ophthalmol. Basel, Karger, 2009, vol 44, pp 37–45

Glial Cell Changes of the Human Retina in Proliferative Vitreoretinopathy

Louisa Wickham · David G. Charteris

Vitreoretinal Unit, Moorfields Eye Hospital, London, UK

Abstract

Retinal detachment initiates a complex series of cellular and molecular changes. Because of the difficulty in obtaining retinal tissue from patients with retinal detachments, most investigations of the cellular changes following retinal detachment have been carried out on animal models. Animal experiments suggest that the cellular response to retinal detachment may be broadly described as; partial dedifferentiation of retinal pigment epithelium cells, proliferation and migration of retinal pigment epithelium cells into the subretinal space, degeneration of the photoreceptor outer segments and synaptic terminals, photoreceptor cell death, structural remodelling of second-order and third-order retinal neurons, proliferation of all non-neuronal cell types within the retina and Müller cell hypertrophy leading to glial scar formation. Each of these stages are considered in turn and related to findings in human proliferative vitreoretinopathy.

Separation of neural retina from the retinal pigment epithelium (RPE) in retinal detachment initiates a complex series of cellular and molecular changes. Evidence that this process might be arrested or reversed following re-apposition of the neural retina with the RPE comes from clinical studies that show good visual outcomes following retinal detachment surgery [1, 2]. However, in some patients the visual outcomes are very poor, particularly in those who develop proliferative vitreoretinopathy (PVR) [3–6], suggesting that remodelling initiated by detachment may continue despite restoration of retinal anatomy. An understanding of cellular changes following retinal detachment together with the wound healing response seen following reattachment is essential in the identification of adjuncts aimed at modifying the development of PVR.

Because of the difficulty in obtaining retinal tissue from patients with retinal detachments, most investigations of the cellular changes following retinal detachment have been carried out on animal models, particularly cat and rabbit. The feline retina is rod dominant, has an intraretinal circulation that is excluded from the photoreceptor layer and a choroidal circulation that supplies the photoreceptor layer. The rabbit retina is also rod dominant but has no intraretinal vasculature with the inner retina being supplied by vessels that lie on the vitreal surface. The rabbit retina is more difficult to conduct experiments on as the rabbit eye is a smaller and the retina tends to degenerate very rapidly following detachment. The study of animal retinal detachment through controlled experiments has provided insight into the cellular effects of different stages of retinal detachment. Broadly these may be described as [7]:

1 Partial dedifferentiation of RPE cells.
2 Proliferation and migration of RPE cells into the subretinal space.
3 Degeneration of the photoreceptor outer segments and synaptic terminals.
4 Photoreceptor cell death.
5 Structural remodelling of second-order and third-order retinal neurons.
6 Proliferation of all non-neuronal cell types within the retina.
7 Müller cell hypertrophy leading to glial scar formation.

Partial Dedifferentiation of RPE Cells

The RPE is a polarised monolayer of neuroepithelial cells that rest on Bruch's membrane [8]. Within 24 h of retinal detachment the long microvilli (5–15 μM) that usually ensheath the photoreceptors are replaced by a fringe of microplicae (~5 μM in length) [9]. Within 1–3 days the usually smooth surface of the RPE becomes mounded with protrusions of cytoplasm into the subretinal space and the nucleus becomes displaced to a more apical position.

Proliferation and Migration of RPE Cells into the Subretinal Space

In the feline model, experiments using ^{3}H-thymidine have shown that within 72 h of retinal detachment the RPE has begun to proliferate and may be observed as areas of hyperplasia within the RPE monolayer, particularly at the junction of attached and detached retina [9, 10]. RPE proliferation is confined to areas of retinal detachment suggesting that the interaction between RPE and neural retinal has an antimitotic effect [9]. RPE proliferation and reversal of cellular polarity may be responsible for affecting regeneration of photoreceptor outer segments following retinal reattachment [11]. In reattached cat retina, areas where the apposition of the retina with the apical surface of the RPE was retained and photoreceptors were quick to regenerate; however, in areas where RPE and Müller cells had proliferated into the sub-retinal space, photoreceptor recovery was poor or nonexistent [12]. Further, to establish a functional relationship between photoreceptors and the RPE, the interaction between the apical surface of the RPE and photoreceptor outer segments must be restored to allow the transport of retinoids vital for the visual cycle [7]. RPE cells with rounded apical surfaces or cells with reversed polarity bear little resemblance to normal RPE and this is likely to influence photoreceptor recovery.

The subretinal space is usually free of cells; however, within 24 h of retinal detachment a number of cell types, e.g. polymorphonuclear neutrophils and monocytes, migrate into this space from the choroidal and retinal capillaries [9]. Free RPE cells are also seen in the subretinal space within 72 h of retinal detachment and frequently contain outer segment fragments, indicating that they may play a role in phagocytosis of cellular debris [9, 13, 14].

Degeneration of the Photoreceptor Outer Segments and Synaptic Terminals

Within 12 h of experimental retinal detachment, photoreceptor outer segments show evidence of structural damage [9]. Initially, the distal end of the outer segment becomes vacuolated or distorted and by 24–72 h all rod and cone outer segments are significantly shorter and distorted with disorientated discs [15]. Outer segment debris is shed into the subretinal space where it is phagocytosed by RPE cells and macrophages which have migrated into the area.

Although retinal detachment interrupts the process of disc production and shedding, outer segment specific proteins continue to be produced but localise to abnormal cellular locations. Opsin, normally concentrated in the outer segment, begins to accumulate in the plasma membrane vitread to the outer segment within a day of experimental retinal detachment (fig. 2) [15]. Peripherin/rds, another outer segment protein specific to the disc rims, is also redistributed and begins to appear in cytoplasmic vesicles [16]. Cone outer segment proteins appear to be more susceptible to damage, with redistributed cone opsins persisting for just 1 week following retinal detachment, after which they are only observed in occasional cells [17].

Inner segments of photoreceptors begin to show signs of degeneration 1–3 days following retinal detachment such as swelling, disruption and loss of mitochondria, disruption of the rough endoplasmic reticulum and Golgi apparatus and an overall reduction in size [7]. The connecting cilium, a structure essential for the production of the outer segment should retinal reattachment occur, is retained even in severely affected inner segments [9].

Photoreceptor Cell Death

In the feline model, photoreceptor cell bodies and synaptic terminals show extensive vacuolisation, degeneration of mitochondria and disorganisation of the microtubules and actin filaments [7]. Degeneration of photoreceptors is variable and areas of extensive degeneration coexist with areas of relatively intact photoreceptors [15].

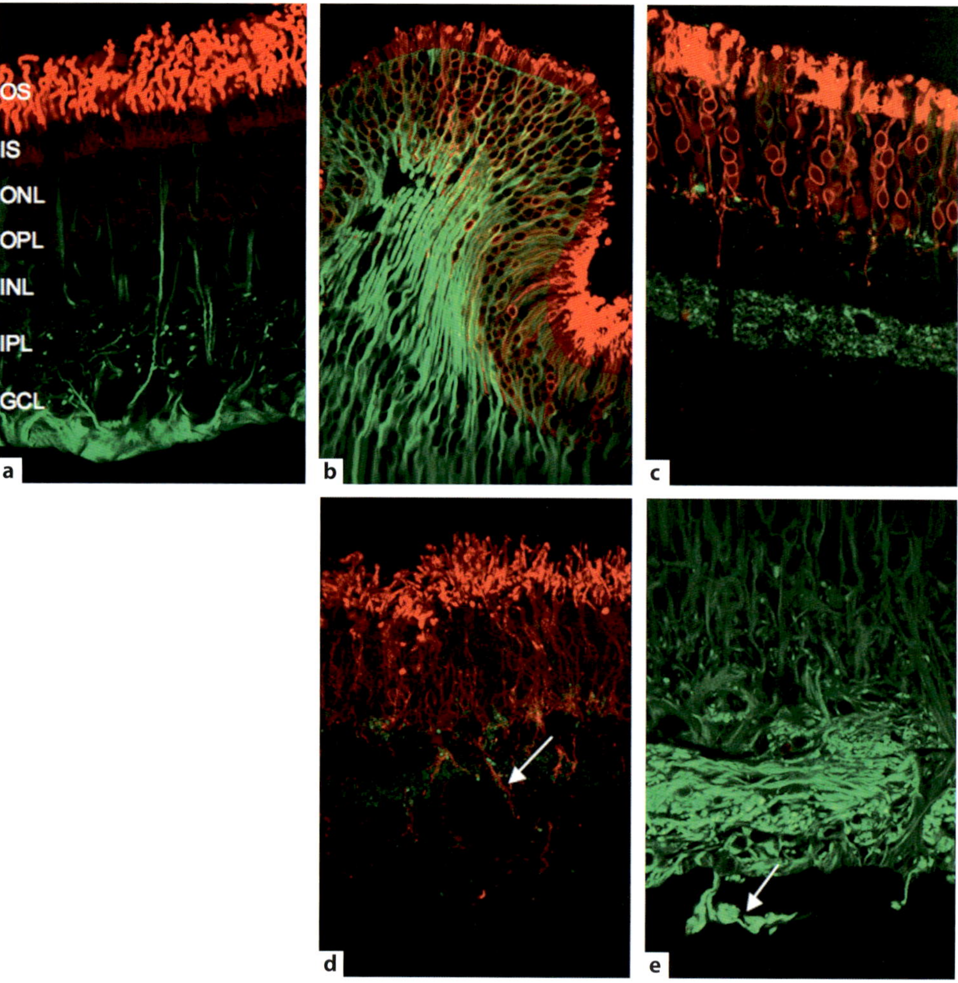

Fig. 1. Double-label laser scanning confocal images of human retina. **a** Normal retina showing anti-rod opsin staining (red) of photoreceptor outer segments and GFAP (green) staining of glial cells. **b** Retinectomy specimen of a patient with grade C PVR. Anti-rod opsin staining (red) of photoreceptor outer segments and redistribution to cell body. GFAP (green). **c** Higher magnification section (Anti rod-opsin – red, synaptophysin green). **d** Beaded neurites extend towards the inner retina (arrow). Staining with synaptophysin (green) demonstrates that some of the neurite varicosities contain synaptic vesicle protein. **e** Müller cell breaching the ILM (arrow) and extending into epiretinal glial tissue. GFAP (green).

Cook et al. [18] demonstrated that cell death via the apoptotic pathway peaks three days following retinal detachment and then continues at low levels for as long as the retina is detached. More recently, photoreceptor apoptosis has also been demonstrated in human retinal specimens, with a peak at 2 days following retinal detachment [19]. In the feline model, approximately 80% of the photoreceptors are lost within 90 days following retinal detachment [20]. Rod cell bodies appear to degenerate quicker than cones following retinal detachment [21], and following cell death photoreceptors are extruded into the subretinal space where they are phagocytosed.

Structural Remodelling of Second-Order and Third-Order Retinal Neurons

Photoreceptor cell bodies extend processes into the outer plexiform layer where they form synapses with second order neurones. Rods and cones have characteristic synaptic terminals called spherules and pedicles respectively. Synaptophysin labelling 3 days following retinal detachment shows many synaptic terminals retracted towards the photoreceptor cell body, with migration of some synaptic structures into the outer nuclear layer from the outer plexiform layer [22]. In the same way that cones appear to degenerate slower than rods, cone pedicles appear to survive longer following retinal detachment than spherules [20].

Changes in the synaptic terminals are accompanied by the growth of processes from the rod bipolar cells and horizontal cells into the outer nuclear layer [22]. Ganglion cells also become reactive, and begin to re-express GAP 43, a protein expressed early in cell body development for the formation of synaptic connections between ganglion cell axons and the brain [23, 24]. Horizontal and ganglion cells show dramatic and extensive remodelling, growing neurites that appear to be attracted to Müller cells and extend into scars both in the subretinal space and on the vitreal surface [23, 25]. By contrast, rod bipolar cell dendrite growth is less aggressive, being targeted towards the retracted terminals of the rod photoreceptors.

Proliferation of Non-Neuronal Cell Types

Within 24 h of retinal detachment all non-neuronal cell types (astrocytes, Müller cells, pericytes, capillary endothelial cells and microglia) display signs of proliferation [26, 27]. Cellular proliferation appears to peak within 3–4 days of retinal detachment after which it slowly reverts to normal levels over a number of weeks [26, 27].

Müller Cell Hypertrophy Leading to Glial Scar Formation

Changes in the Müller cell are seen within 1 day of retinal detachment, including changes in protein expression and early growth of Müller cell processes [28]. Within 3 days, Müller cell bodies have migrated to the outer nuclear and outer plexiform layers into the spaces left by dying photoreceptors, and processes begin to extend into the subretinal space through localised disruptions in the outer limiting membrane. Following retinal detachment, Müller cell processes within the retina show greatly increased expression of both vimentin and glial fibrillary acidic protein (GFAP), whereas segments of the Müller cell processes that extend through the outer limiting membrane into the subretinal space preferentially express vimentin [28, 29]. These processes appear to penetrate the outer limiting membrane adjacent to cone photoreceptors, often growing for long distances on the photoreceptor border and later forming multi-layered scars [28]. Following retinal reattachment, Müller cell processes within the subretinal space appear to inhibit the regeneration of photoreceptors and may therefore be influential in determining visual recovery [30].

Retinal Recovery and the Development of PVR following Re-attachment

Retinal recovery following retinal attachment relies on re-establishing the cell-to-cell contact of RPE cells and photoreceptors. This involves redifferentiation of the RPE apical surface, re-ensheathment of the regenerating outer segments, and probably resynthesis of inter-photoreceptor matrix components. In the feline model, retinal reattachment within 1–3 days is very effective at reversing the cellular changes induced by retinal detachment [9, 31]. Following reattachment outer segments recover to approximately 70% of their length, photoreceptor apoptosis is arrested and cellular proliferation is reduced [32]. Restoration of retinal anatomy following detachment is not uniform. The morphological appearance has been described as a 'patchwork', with areas of variability in outer segment length and levels of protein expression both within areas of retina and between different retinas [15, 32].

In areas where Müller cell hypertrophy has led to the formation of a subretinal scar, photoreceptors only show limited signs of recovery with no outer segment regeneration [30]. Moreover, reattachment appears to induce growth of Müller cell processes into the vitreous which form epiretinal membranes and act as a substrate for ganglion cell growth and neurite extension and also appear to induce growth of rod axons into the inner retina [28, 30].

Cellular Changes in Human PVR

Human correlates of animal experiments exist. Retinal tissue removed from post mortem specimens and from patients undergoing retinal detachment surgery has demonstrated changes similar to those seen in animal models [33–36]. In animal studies, retinal detachments are induced in a controlled manner using a micropipette to inject fluid between the neural retina and RPE. This differs from human studies where the precise onset of retinal detachment can be difficult to determine and may vary within the eye according to its rate of progression. In addition, the process of acute retinal tear formation can results in the liberation of mitotically active RPE cells into the subretinal and preretinal spaces. These RPE cells may subsequently contribute to the formation of complex epiretinal membranes seen in human PVR [37]. A case report of an acute retinal detachment showed that retinal tissue adjacent to the retinal tear demonstrated more advanced degenerative changes when compared with that from areas more distant from the breaks [36].

When retinal detachment is complicated by PVR samples obtained from patients undergoing retinectomy provide an insight into the pathology in more chronic stages of retinal detachment. In a study of 16 retinectomy specimens taken from patients with PVR, Sethi et al. [34] demonstrated that the response to prolonged retinal detachment with PVR in humans was similar to that observed in chronic detachment animal models. In humans, as in animal models, photoreceptors were degenerate and intracellular redistribution of opsin proteins to the plasma membrane was observed. In cones, labelling with anti M/L cone opsin showed degenerate outer segments and faint staining of swollen inner segments and in severe PVR staining of cone opsins was absent. Rod synaptic terminals showed remodelling with extension of rod bipolar cell dendrites and horizontal cell processes into the outer nuclear layer.

There was also upregulation of neurofilament and growth-associated protein-43 (GAP-43) expression in large ganglion cells with neurite sprouting. All retinectomy specimens showed a marked upregulation of Müller cell and astrocyte expression of GFAP and vimentin with areas of increased glial tissue replacing degenerated retinal neurons [34]. In some sections Müller cells breached the ILM and extended onto the retinal surface where they formed a component of epiretinal membrane [38]. The presence of neurites in sub- and epiretinal membranes has also been demonstrated in retinal specimens taken at the time of PVR surgery [39].

Close correlation of human and animal models of retinal detachment support the use of animal models in the characterisation of glial cell changes in PVR. Further, animal models are a valuable resource allowing investigation of the effects of potential adjuncts in this complex process.

References

1 Ah-Fat FG, Sharma MC, Majid MA: Trends in vitreoretinal surgery at a tertiary referral centre: 1987 to 1996. Br J Ophthalmol 1999;83:396–398.

2 Wickham L, Bunce C, Wong D, McGurn D, Charteris DG: Randomized controlled trial of combined 5-fluorouracil and low-molecular-weight heparin in the management of unselected rhegmatogenous retinal detachments undergoing primary vitrectomy. Ophthalmology 2007;114: 698–704.

3 Silicone Study Group: Vitrectomy with silicone oil or perfluoropropane gas in eyes with severe proliferative vitreoretinopathy: results of a randomized clinical trial. Silicone Study Report 2. Arch Ophthalmol 1992;110:780–792.

4 Silicone Study Group: Vitrectomy with silicone oil or sulfur hexafluoride gas in eyes with severe proliferative vitreoretinopathy: results of a randomized clinical trial. Silicone Study Report 1, Arch Ophthalmol 1992;110:770–779.

5 Lewis H, Aaberg T: Causes of failure after repeat vitreoretinal surgery for recurrent proliferative vitreoretinopathy. Am J Ophthalmol 1991;111:15–19.

6 Lewis H, Aaberg TM, Abrams GW: Causes of failure after initial vitreoretinal surgery for severe proliferative vitreoretinopathy, Am J Ophthalmol 1991;111: 8–14.

7 Fisher SK, Anderson DH: Cellular effects of detachment on the neural retina and the retinal pigment epithelium; in Wilkinson CP (ed): Retina. St Louis, Mosby, 2001, pp 1961–1986.

8 Campochiaro PA, Jerdan JA, Glaser BM: The extracellular matrix of human retinal pigment epithelial cells in vivo and its synthesis in vitro. Invest Ophthalmol Vis Sci 1986;27:1615–1621.

9 Anderson DH, Stern WH, Fisher SK, Erickson PA, Borgula GA: Retinal detachment in the cat: the pigment epithelial-photoreceptor interface. Invest Ophthalmol Vis Sci 1983;24:906–926.

10 Anderson DH, Stern WH, Fisher SK, Erickson PA, Borgula GA: The onset of pigment epithelial proliferation after retinal detachment. Invest Ophthalmol Vis Sci 1981;21:10–16.

11 Guerin CJ, Anderson DH, Fariss RN, Fisher SK: Retinal reattachment of the primate macula: photoreceptor recovery after short-term retinal detachment. Invest Ophthalmol Vis Sci 1989;30: 1708–1725.

12 Stern WH, Anderson DH, Fisher SK, Borgula GA, Ericson PA: Anatomical recovery following retinal detachment: clinicopathological correlations. Aust J Ophthalmol 1981;9:143–154.

13 Johnson NF, Foulds WS: Observations on the retinal pigment epithelium and retinal macrophages in experimental retinal detachment, Br J Ophthalmol 1977;61:564–572.

14 Elner VM, Schaffner T, Taylor K, Glagov S: Immunophagocytic properties of retinal pigment epithelium cells. Science 1981;211:74–76.

15 Lewis GP, Sethi CS, Linberg KA, Charteris DG, Fisher SK: Experimental retinal reattachment: a new perspective. Mol Neurobiol 2003;28:159–175.

16 Fariss RN, Molday RS, Fisher SK, Matsumoto B: Evidence from normal and degenerating photoreceptors that two outer segment integral membrane proteins have separate transport pathways. J Comp Neurol 1997;387:148–156.

17 Mervin K, Valter K, Maslim J, Lewis GP, Fisher SK, Stone J: Limiting photoreceptor death and deconstruction during experimental retinal detachment: the value of oxygen supplementation. Am J Ophthalmol 1999;128:155–164.

18 Cook B, Lewis GP, Fisher SK, Adler R: Apoptotic photoreceptor degeneration in experimental retinal detachment. Invest Ophthalmol Vis Sci 1995;36:990–996.

19 Arroyo JG, Yang L, Bula D, Chen DF: Photoreceptor apoptosis in human retinal detachment. Am J Ophthalmol 2005;139:605–610.

20 Erickson PA, Fisher SK, Anderson DH, Stern W, Borgula GA: Retinal detachment in the cat: the outer nuclear and outer plexiform layers. Invest Ophthalmol Vis Sci 1983;24:927–942.

21 Rex TS, Fariss RN, Lewis GP, Linberg KA, Sokal I, Fisher SK: A survey of molecular expression by photoreceptors after experimental retinal detachment. Invest Ophthalmol Vis Sci 2002;43:1234–1247.

22 Lewis GP, Linberg KA, Fisher SK: Neurite outgrowth from bipolar and horizontal cells after experimental retinal detachment. Invest Ophthalmol Vis Sci 1998;39:424–434.

23 Coblentz FE, Radeke MJ, Lewis GP, Fisher SK: Evidence that ganglion cells react to retinal detachment. Exp Eye Res 2003;76:333–342.

24 Dijk F, Bergen AA, Kamphuis W: GAP-43 expression is upregulated in retinal ganglion cells after ischemia/reperfusion-induced damage. Exp Eye Res 2007;84:858–867.

25 Fisher SK, Lewis GP, Linberg KA, Verardo M: Cellular remodelling in mammalian retina: results from studies of experimental retinal detachment. Prog Retin Eye Res 2005;24:395–431.

26 Fisher SK, Erickson PA, Lewis GP, Anderson DH: Intraretinal proliferation induced by retinal detachment. Invest Ophthalmol Vis Sci 1991;32:1739–1748.

27 Geller SF, Lewis GP, Anderson DH, Fisher SK: Use of MB1 antibody for detecting proliferating cells in the retina. Invest Ophthalmol Vis Sci 1995;36:737–744.

28 Lewis GP, Fisher SK: Upregulation of glial fibrillary acidic protein in response to retinal injury: its potential role in glial remodelling and a comparison of vimentin expression. Int Rev Cytol 2003;230:263–290.

29 Lewis GP, Erickson PA, Guerin CJ, Anderson DH, Fisher SK: Changes in the expression of specific Muller cell proteins during long-term retinal detachment. Exp Eye Res 1989;49:93–111.

30 Fisher SK, Lewis GP: Mueller cell and neuronal remodelling in retinal detachment and reattachment and their potential consequences for visual recovery: a review and reconsideration of recent data. Vision Res 2003;43:887–897.

31 Lewis GP, Charteris DG, Sethi CS, Leitner WP, Linberg KA, Fisher SK: The ability of rapid retinal reattachment to stop or reverse the cellular and molecular events initiated by detachment. Invest Ophthalmol Vis Sci 2002;43:2412–2420.

32 Lewis GP, Charteris DG, Sethi CS, Fisher SK: Animal models of retinal detachment and reattachment: identifying cellular events that may affect visual recovery. Eye 2002;16:375–387.

33 Chang CJ, Lai WW, Edward DP, Tso MO: Apoptotic photoreceptor cell death after traumatic retinal detachment in humans. Arch Ophthalmol 1995; 113:880–886.

34 Sethi CS, Lewis GP, Fisher SK: Glial remodelling and neuronal plasticity in human retinal detachment with proliferative vitreoretinopathy. Invest Ophthalmol Vis Sci 2005;46:329–342.
35 Wilson DJ, Green WR: Histopathological study of the effect of retinal detachment surgery on 49 eyes obtained post mortem. Am J Ophthalmol 1987;103: 167–179.
36 Wickham L, Sethi CS, Lewis GP, Fisher SK, McLeod DC, Charteris DG: Glial and neural response in short-term human retinal detachment. Arch Ophthalmol 2006;124:1779–1782.
37 Charteris DG: Proliferative vitreoretinopathy: pathobiology, surgical management, and adjunctive treatment. Br J Ophthalmol 1995;79:953–960.
38 Lewis GP, Betts KE, Sethi CS, Charteris DG, Lesnik-Oberstein SY, Avery RL, Fisher SK: Identification of ganglion cell neurites in human subretinal and epiretinal membranes. Br J Ophthalmol 2007;91: 1234–1238.
39 Lewis GP, Betts KE, Sethi CS, Charteris DG, Lesnik-Oberstein SY, Avery RL, Fisher SK: Identification of ganglion cell neurites in sub- and epiretinal membranes. Br J Ophthalmol 2007;91:1234–1238.

Louisa Wickham
VR Research Fellow, Moorfields Eye Hospital
City Road
London EC1V 2PD (UK)
Tel. 0207 5662285, Fax 0207 6086925, E-mail louisa.wickham@moorfields.nhs.uk

Gandorfer A (ed): Pharmacology and Vitreoretinal Surgery.
Dev Ophthalmol. Basel, Karger, 2009, vol 44, pp 46–55

Alkylphosphocholines: A New Approach to Inhibit Cell Proliferation in Proliferative Vitreoretinopathy

Kirsten H. Eibl[a] · Steven K. Fisher[b] · Geoffrey P. Lewis[b]

[a]University Eye Hospital, Ludwig-Maximilians-University, Munich, Germany; [b]Neuroscience Research Institute, University of California, Santa Barbara, Calif., USA

Abstract

Proliferative vitreoretinopathy represents the major complication in retinal detachment surgery and occurs in about 5–15% of cases resulting in a significant loss of vision despite multiple surgical procedures. Although successful anatomical reattachment is usually achieved, the reduction in central vision often remains permanent due to the intraretinal changes induced by retinal detachment and the subsequent proliferative response within the retina. Retinal Müller glial cells play a pivotal role in this process together with retinal pigment epithelial cells which are dispersed in the vitreous and stimulated by growth factors and serum in the vitreous after the breakdown of the blood-retinal barrier.

Alkylphosphocholines (APCs) are effective inhibitors of human ocular cell proliferation and currently in clinical use for the treatment of protozoan and malignant diseases. Previously, we have demonstrated that APCs can inhibit human retinal Müller glial cell proliferation and attenuate the hypoxia-induced actin filament expression in these cells. Moreover, retinal pigment epithelial (RPE) cell spreading and migration as well as proliferation and cell-mediated membrane contraction are effectively inhibited by APCs at nontoxic concentrations in vitro. The first in vivo toxicity studies in the rat eye also did not display any toxic effect of APCs to retinal tissue 7 days after intravitreal administration. Finally, in a well-established in vivo model for PVR, we were able to demonstrate that APCs can dramatically reduce intraretinal Müller glial cell proliferation induced by experimental retinal detachment in nontoxic concentrations. Thus, APCs are promising, novel pharmacologic substances that may be a useful adjunct to reattachment surgery by reducing the intraretinal proliferation induced by detachment and hence reduce the incidence of PVR.

Cell Biology of Retinal Detachment and Proliferative Vitreoretinopathy

Following retinal detachment, a cascade of events are set in motion: photoreceptor outer segments degenerate, some photoreceptors die by way of apoptosis [1] and synaptic terminals undergo significant remodeling. In response to changes in photoreceptors, in particular the loss of synaptic connections, second-order neurons extend neurites into the outer retina [2]. Ultimately, the ganglion cells respond to detachment by also elaborating new processes throughout the retina [3]. While loss of photoreceptors and 'downstream' neuronal changes most certainly contribute to a loss in visual acuity, non-neuronal cells, in particular the glial cells, are also thought to play a role in the continued reduction in vision following retinal reattachment surgery and contribute to the subsequent proliferative response that occurs within the retina [4, 5]. Following experimental retinal detachment, it has been shown that all non-neuronal cell types are stimulated to divide [6, 7]. These cell types include Müller cells, astrocytes, RPE cells, microglia and macrophages; however, the Müller glial cell appears to be the cell type undergoing the highest level of cell division. Indeed, Müller cell proliferation has been implicated in the pathogenesis of various human retinal diseases including retinal degeneration [8], retinal detachment and PVR [9], proliferative diabetic retinopathy (PDR) [10] and epiretinal membrane formation in idiopathic surface wrinkling maculopathy [11–13]. Moreover, Müller cells are thought to play a role in the degenerative events occurring in attached retinal regions within the detached eye [5, 14].

Intraretinal glial cell proliferation is one of the earliest responses to retinal detachment [6, 7]. It is accompanied by rapid glial hypertrophy and upregulation of intermediate filament proteins [15]. These events most likely contribute to the growth of these cells onto either the sub- or epiretinal surface. Indeed, Müller cells have been identified in both human retinal tissue [9, 11–13] and experimental animal models of retinal detachment [6, 8]. Importantly, the presence of these glial 'membranes' can have devastating consequences for the return of good vision. With subretinal growth, termed subretinal fibrosis, glial cells can prevent the regrowth of photoreceptor outer segments following reattachment surgery [16] while epiretinal growth can cause re-detachment of the retina [17]. The fact that these events occur rapidly soon after detachment indicates that treatment with antiproliferative agents should also begin at this time. In addition, since it has been shown that proliferation continues following retinal reattachment, albeit at low levels [18], successful inhibition of membrane formation may require not only early but also long-term delivery of the agent.

The stimulus for initiating the proliferation of cells is unknown; however, due to the breakdown of the blood-retinal barrier after retinal detachment, there is certainly an increased amount of serum and growth factors present in the vitreous cavity that may promote proliferation of Müller glial and RPE cells [19, 20]. Indeed, intravitreal injection of basic fibroblast growth factor into a normal feline eye, has been shown to induce proliferation of non-neuronal cells, upregulate intermediate filament proteins, and stimulate the formation of epiretinal membranes [21].

Another factor that is most certainly involved in the response of the retina following detachment is hypoxia of the outer retina as the neural retina becomes distanced from the choriocapillaris. Hypoxia is known to cause significant changes in the architecture of the actin cytoskeleton and thus plays a pivotal role in complex cellular events like migration and proliferation [22]. Moreover, treatment of animals with retinal detachment by hyperoxia (70% inhaled O_2) results in reduced photoreceptor deconstruction and fewer dividing Müller cells [23]. Finally, while PVR is generally considered a 'disease' of proliferation, cellular hypertrophy is also most certainly involved. Thus, the crucial pathophysiologic events to be influenced by a novel intravitreal pharmacotherapy are first, the serum/growth factor-induced proliferation/hypertrophy of Müller glial and RPE cells, and second the hypoxia-induced glial changes with deconstruction of photoreceptors.

Current Therapeutic Concepts

Surgical intervention in itself is insufficient to control the cellular responses of PVR and, therefore, a pharmacological concept is warranted as an adjuvant therapy in combination with complex retinal reattachment surgery. Current pharmacologic intervention focuses primarily on antiproliferative and anti-inflammatory agents to prevent PVR [24]. A number of antiproliferative pharmacologic substances such as colchicine, daunomycin and 5-fluouracil have been tested for their effects on human retinal glial cells in vitro [25]. However, these substances are currently not part of any routine clinical treatment for PVR. One of the most promising candidates, 5-fluouracil, combined with low-molecular heparin, has recently been tested in a large clinical trial; however, it proved to be ineffective in reducing the incidence of PVR [26, 27]. A number of factors most likely contribute to the failure of studies such as these, one of which certainly involves choosing the optimal treatment regime. Based on animal studies, the ideal treatment would involve early administration of the agent, perhaps at the time of reattachment surgery in high-risk cases, combined with sustained delivery without causing toxicity to the retina. We feel that alkylphosphocholines (APCs), encapsulated in liposomes meet these criteria.

Alkylphosphocholines

Alkylphosphocholines (APCs) are synthetic phospholipid derivatives (fig. 1) and known as effective inhibitors of cellular proliferation. Clinically, they are currently applied for the treatment of protozoan [28] and malignant diseases [29]. In addition, APCs have been shown to inhibit human ocular cell proliferation which may have therapeutic implications for a number of proliferative ocular conditions including

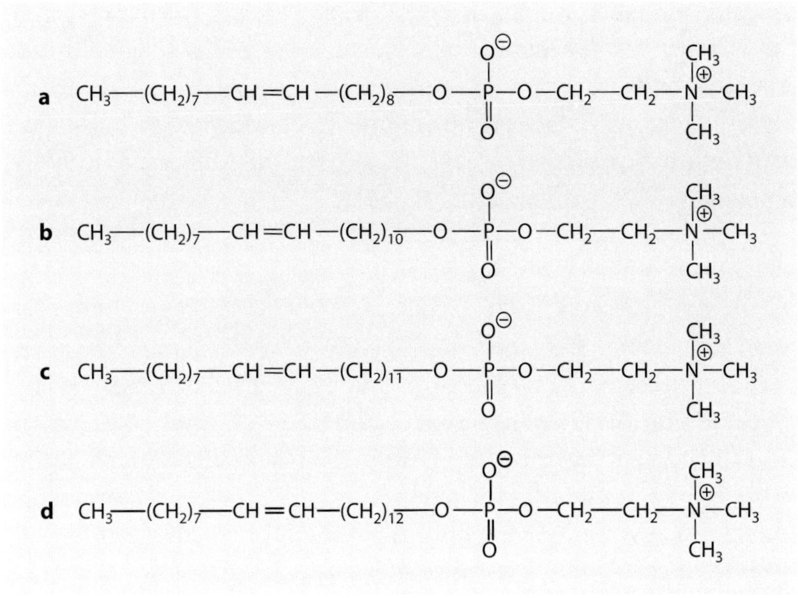

a $CH_3{-}(CH_2)_7{-}CH{=}CH{-}(CH_2)_8{-}O{-}P(O^{\ominus})(=O){-}O{-}CH_2{-}CH_2{-}N^{\oplus}(CH_3)_3$

b $CH_3{-}(CH_2)_7{-}CH{=}CH{-}(CH_2)_{10}{-}O{-}P(O^{\ominus})(=O){-}O{-}CH_2{-}CH_2{-}N^{\oplus}(CH_3)_3$

c $CH_3{-}(CH_2)_7{-}CH{=}CH{-}(CH_2)_{11}{-}O{-}P(O^{\ominus})(=O){-}O{-}CH_2{-}CH_2{-}N^{\oplus}(CH_3)_3$

d $CH_3{-}(CH_2)_7{-}CH{=}CH{-}(CH_2)_{12}{-}O{-}P(O^{\ominus})(=O){-}O{-}CH_2{-}CH_2{-}N^{\oplus}(CH_3)_3$

Fig. 1. Chemical structure of the alkylphosphocholines oleyl-phosphocholine (C18:1-PC, **a**), (Z)-10-eicosenyl-phosphocholine (C20:1-PC, **b**), (Z)-12-heneicosenyl-phosphocholine (C21:1-PC, **c**), and erucyl-phosphocholine (C22:1-PC, **d**) applied in ophthalmic research.

those secondary to retinal detachment [30] as well as the scar formation of Tenon fibroblasts after glaucoma filtration surgery [31].

Their mechanism of action involves binding to the membrane-bound G-protein PKC (protein kinase C) which is part of a major intracellular second-messenger systems that regulates cell attachment, spreading, migration and proliferation. In ocular cells (Tenon fibroblasts and retinal pigment epithelial cells) and in the leukemia cell line HL60, APCs are effective inhibitors of PKC [31–33]. Also, PKC has been shown to be associated with a downregulation of focal adhesion kinase (FAK) [34]. FAK is one of the key cytoplasmic tyrosine kinases and acts as a potential integrin effector. A high proliferation rate is associated with an increased FAK activity in malignant glioma cells [35]. Moreover, FAK directs signaling molecules like phosphatidylinositol 3-kinase-Akt/PKB (PI3K) to focal contacts between cells and decides if cells migrate or proliferate [36]. Since APCs have been reported to inhibit the PI3K pathway in epithelial carcinoma cell lines [37], actin organization and focal contact dynamics which are in part determined by a PKC [38] and a FAK-directed signaling cascade might be influenced by APCs via both pathways.

Previously, we have been able to demonstrate that APCs can inhibit RPE cell spreading and migration [39] as well as proliferation and cell-mediated membrane contraction [32] at nontoxic concentrations in vitro. Moreover, the first in vivo toxicity studies in the rat eye did not display any toxic effect of APCs to retinal tissue 7 days after intravitreal administration [40]. Recent in vitro [41] and in vivo [30] studies examining the effect of alkylphosphocholines on retinal Müller glial cells have also been performed to further elucidate their potential therapeutic role in retinal proliferative diseases. Thus, APCs are promising, novel pharmacologic substances which have been found to be effective in controlling both the Müller cell [36] and RPE-related component of PVR pathogenesis [32, 34].

In Vitro Effects

To examine the in vitro effects of APCs on retinal Müller glial cells, primary Müller glial cells were prepared from Long-Evans rats in accordance with applicable German laws and the ARVO Statement for the Use of Animals in Ophthalmic and Vision Research. Isolated cells were routinely evaluated by immunofluorescence microscopy using a monoclonal antibody against vimentin. Also, the spontaneously immortalized human Müller cell line, MIO-M1, was used to evaluate the role of the alkylphosphocholines oleyl-phosphocholine (C18:1-PC; fig. 1a) and erucyl-phosphocholine (C22:1-PC; fig. 1d) on cell proliferation and F-actin stress fiber expression under normoxic and hypoxic conditions. To determine the levels of Müller glial cell proliferation, the incorporation of 5-bromo-2-deoxyuridine (BrdU) was measured after assessment of toxicity by the trypan blue exclusion assay. For detection of F-actin stress fibers, the rhodamine-phalloidin staining protocol was applied.

With this in vitro study, we could demonstrate that C18:1-PC and C22:1-PC are able to inhibit the proliferation of primary rat retinal Müller glial cells (fig. 2a) as well as of the immortalized human retinal Müller glial cell line MIO-M1 (fig. 2b) in a dose-dependent manner. Effective concentrations were non-toxic and able to attenuate the content of F-actin stress fibers in these cells which were induced by hypoxia, a key feature in the pathogenesis of retinal diseases. Therefore, in conclusion, this study provides further evidence that APCs might offer a novel treatment strategy for the prevention of PVR: first, they inhibit Müller glial cell proliferation and second, they are able to diminish the hypoxia-related upregulation of F-actin stress fibers in vitro [41].

In Vivo Studies

A subsequent evaluation of the results obtained in vitro has been performed in a well established in vivo model for PVR, the rabbit eye, where APCs have been found to be effective inhibitors of intraretinal proliferation after experimental retinal detachment [30].

Retinal detachments were created in adult New Zealand Red pigmented rabbits by infusing a solution of sodium hyaluronate (0.25% in balanced salt solution; BSS)

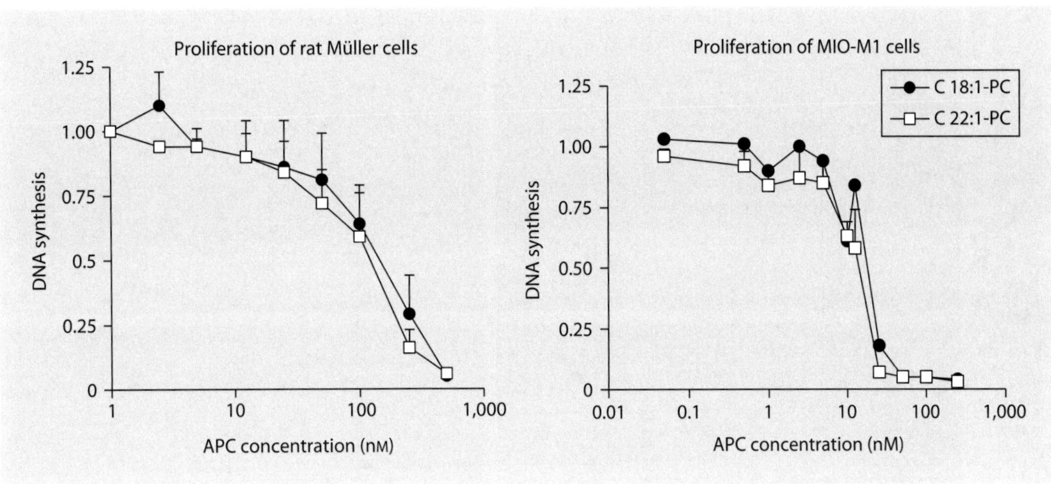

Fig. 2. C18:1-PC and C22:1-PC inhibit proliferation of retinal Müller glial cells. Primary rat Müller glial cells (**a**) or MIO-M1 cells (**b**) were cultured in 96-well microplates and incubated with different concentrations of the APCs. Proliferation of cells was assessed by measurement of BrdU incorporation ($p < 0.05$ for all concentrations >25 nM).

via a glass pipette between the neural retina and RPE. Two experimental conditions were tested: (1) whether APC treatment was more effective given at day 1 or day 2 after detachment, and (2) whether liposome-bound APC was more effective than the free drug. Control eyes were injected with balanced salt solution on day 1. In treated eyes, 100 μM of APC in 50 μl BSS was injected intravitreally via a 30-gauge needle. On day 3 after detachment, the animals were injected with 10 μg BrdU. The eyes were bisected and half of each eye, containing both attached and detached regions, was fixed in 4% paraformaldehyde for immunocytochemical analysis. Anti-BrdU was used to detect dividing cells, antivimentin to determine the extent of Müller cell hypertrophy and isolectin B4, *Griffonia simplicifolia* to label microglia and macrophages. The other half was placed in 1% paraformaldehyde plus 1% glutaraldehyde in phosphate buffer for light and electron microscopic analysis to determine potential toxic effects of APC. All experimental procedures conformed to the ARVO Statement for the Use of Animals in Ophthalmic and Vision Research and the guidelines of the Animal Resource Center of the University of California, Santa Barbara, Calif., USA.

In control, nondetached retina, no anti-BrdU-labeled cells were detected (fig. 3a). In control detachments, Müller cells represented the largest subgroup of BrdU-labeled cells (fig. 3b, red) contributing to the massive intraretinal scar formation. The isolectin B4-labeled microglia within the retina formed the second largest subgroup and the isolectin B4-labeled macrophages in the subretinal space formed the third (fig 3a–e, both cell types in blue). In the APC injected eyes, the number of

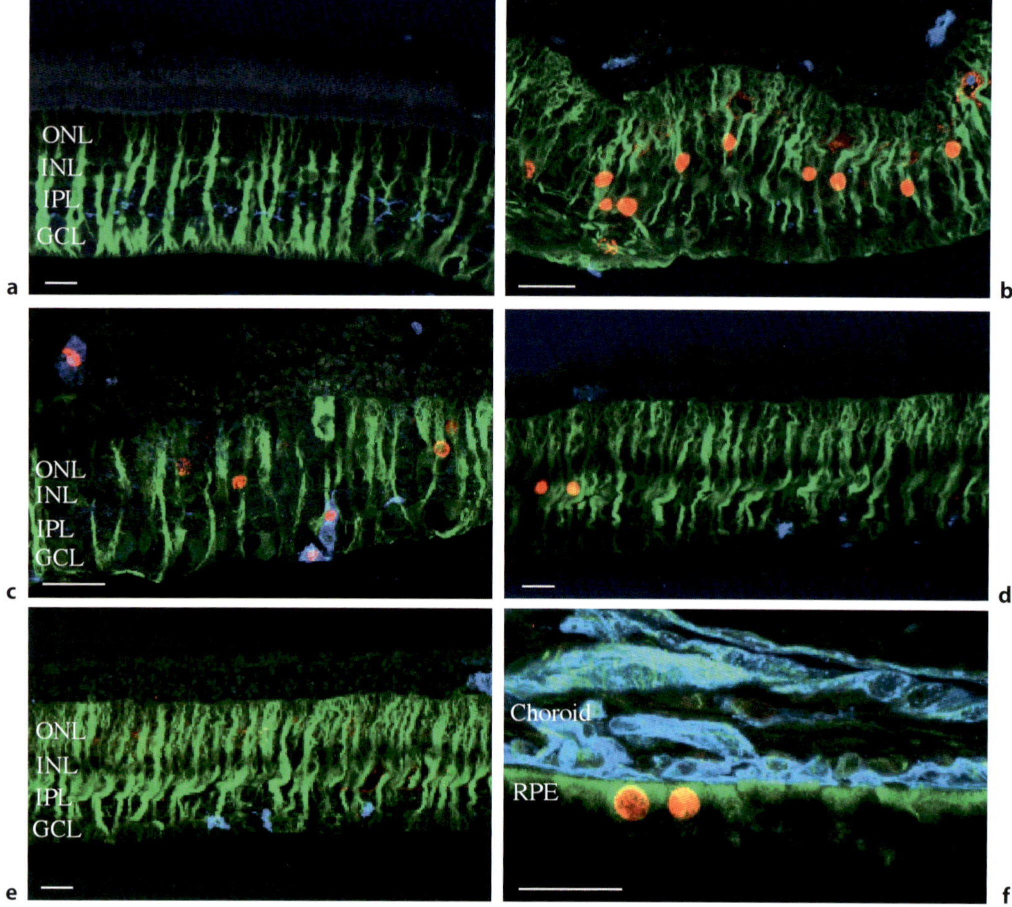

Fig. 3. Laser scanning confocal images of retinal sections labeled with probes to vimentin (green), BrdU (red) and isolectin B4 (blue). **a** Undetached control retina. Anti-vimentin labeled Müller cells extend from the ganglion cell layer (GCL) into the outer nuclear layer (ONL), and isolectin B4 labeled fine microglial cell processes in the inner plexiform layer (IPL). **b** Detached control retina 3 days after detachment and saline injection on day 1. Anti-vimentin labeling increased in Müller cells to the outer limiting membrane (OLM), anti-BrdU labeled dividing cells primarily in the inner nuclear layer (INL), and isolectin B4 labeled microglia that had rounded up and migrated throughout the retina. **c** Three days after retinal detachment and non-liposome-bound APCs injected on day 2. A slight reduction in the number of anti-BrdU-labeled Müller cells was observed; antivimentin and isolectin B4 labeling appeared similar to the control. **d**, **e** Three days after retinal detachment and liposome bound drug injected on day 1. There was a significant decrease in anti-BrdU-labeled Müller cell nuclei although anti-vimentin and isolectin B4 labeling appeared similar to controls. **f** RPE region, 3 days after retinal detachment and saline injection on day 1. A few anti-BrdU labeled cells were observed but this pattern was not significantly different from the drug-treated animals. Scale bars = 20 μm.

BrdU-labeled Müller cells was significantly reduced (fig. 3c–e) without effecting the number of microglia or macrophages. In both control and APC-treated animals, the anti-vimentin labeling increased in Müller cells to the outer limiting membrane and sometimes beyond indicating cellular hypertrophy in response to retinal detachment and hypoxia (fig. 3b, c). Anti-BrdU-labeled Müller cells were observed both in the inner nuclear layer (INL), their normal location in the retina, and in the ONL, as a result of migration into the outer retina (fig. 3b). Numerous anti-BrdU-labeled Müller cells and isolectin B4-positive cells were observed in retinas given free APC on day 2 (fig. 3c), whereas significantly fewer cells incorporated BrdU when the retinas were treated with the liposome-bound APC given on either day 1 (fig. 3d, e) or day 2. There was no evidence of toxicity detected in any of the treatment groups since the retinal morphology at the electron-microscopic level of treated animals appeared the same as in the control eyes [30]. The RPE was also stimulated to divide following detachment, as shown by BrdU incorporation (fig. 3f); however, the number of labeled cells was too low to detect a difference between treated and control animals.

Conclusions/Perspectives

In these recent in vitro and in vivo studies, we have been able to demonstrate the antiproliferative effect of alkylphosphocholines on retinal Müller glial cells at nontoxic concentrations. By using a well-recognized animal model of experimental retinal detachment, we could show the safety and efficacy of APCs for inhibition of Müller cell proliferation after intravitreal injection in the rabbit. The effects on glial cell hypertrophy and expansion outside the retina, however, have yet to be determined. The in vitro assessment of F-actin stress fiber expression and distribution demonstrated a protective effect of APCs on human and rat Müller glial cells under hypoxic cell culture conditions.

The use of alkylphosphocholines may represent an important step towards the development of an adjuvant pharmacologic treatment strategy for PVR and clinical application of these compounds seems feasible. However, long-term toxicity studies in an in vivo model must be performed before APCs can be considered for clinical application.

Grant Support

DOG Research Award 2003 (K.H.E.), NIH grant EY000888 (S.K.F.). Alcon Retina Scholarship (K.H.E.).

References

1 Cook B, Lewis GP, Fisher SK, Adler R: Apoptotic photoreceptor degeneration in experimental retinal detachment. Invest Ophthalmol Vis Sci 1995;36:990–996.

2 Lewis GP, Linberg KA, Fisher SK: Neurite outgrowth from bipolar and horizontal cells following experimental retnal detachment. Invest Ophthalmol Vis Sci 1998;39:424–434.

3 Coblentz FE, Radeke MJ, Lewis GP, Fisher SK: Evidence that ganglion cells react to retinal detachment. Exp Eye Res 2003;76:333–342.

4 Fisher SK, Lewis GP: Müller cell and neuronal remodelling in retinal detachment and reattachment and their potential consequences for visual recovery: a review and reconsideration of recent data. Vis Res 2003;43:887–897.

5 Francke M, Faude F, Pannicke T, Uckermann O, Weick M, Wolburg H, Wiedemann P, Reichenbach A, Uhlmann S, Bringmann A: Glial cell-mediated spread of retinal degeneration during detachment: a hypothesis based upon studies in rabbits. Vision Res 2005;45:2256–2267.

6 Fisher SK, Erickson PA, Lewis GP, Anderson DH: Intraretinal proliferation induced by retinal detachment. Invest Ophthalmol Vis Sci 1991;32:1739–1748.

7 Geller SF, Lewis GP, Anderson DH, Fisher SK: Use of the MIB-1 antibody for detecting proliferating cells in the retina. Invest Ophthalmol Vis Sci 1995;36:737–744.

8 Reichenbach A, Wurm A, Pannicke T, Iandiev I, Wiedemann P, Bringmann A: Müller cells as players in retinal degeneration and edema. Graefes Arch Clin Exp Ophthalmol 2007;245:627–636.

9 Sethi CS, Lewis GP, Fisher SK, Leitner WP, Mann DL, Luthert PJ, Charteris DG. Glial remodeling and neuronal plasticity in human retinal detachment with proliferative vitreoretinopathy. Invest Ophthalmol Vis Sci 2005;46:329–342.

10 Harada C, Harada T, Mitamura Y, Quah HM, Ohtsuka K, Kotake S, Ohno S, Wada K, Takeuchi S, Tanaka K: Diverse NF-kappaB expression in epiretinal membranes after human diabetic retinopathy and proliferative vitreoretinopathy. Mol Vis 2004;10: 31–36.

11 Haritoglou C, Schumann RG, Kampik A, Gandorfer A: Glial cell proliferation under the internal limiting membrane in a patient with cellophane maculopathy. Arch Ophthalmol 2007;125:1301–1302.

12 Kampik A, Green WR, Michels RG, Nase PK: Ultrastructural features of progressive idiopathic epiretinal membrane removed by vitreous surgery. Am J Ophthalmol 1980;90:979–809.

13 Kampik A, Kenyon KR, Michels RG, Green WR, de la Cruz ZC: Epiretinal and vitreous membranes: comparative study of 56 cases. Arch Ophthalmol 1981;99:1445–1454.

14 Faude F, Francke M, Makarov F, Schuck J, Gärtner U, Reichelt W, Wiedemann P, Wolburg H, Reichenbach A: Experimental retinal detachment causes widespread and mutlilayered degeneration in the rabbit retina. J Neurocytol 2001;30:379–390.

15 Fisher SK, Lewis GP, Linberg KA, Verardo MR: Cellular remodeling in mammalial retina: results from studies of experimental retinal detachment. Prog Retin Eye Res 2005;24:395–431.

16 Anderson DH, Guerin CJ, Erickson PA, Stern WH, Fisher SK: Morphological recovery in the reattached retina. Invest Ophthalmol Vis Sci 1986;27:168–183.

17 Charteris DG, Downie J, Aylward GW, Sethi C, Luthert P: Intraretinal and periretinal pathology in anterior proliferative vitreoretinopathy. Graefes Arch Clin Exp Ophthalmol 2007;245:93–100.

18 Lewis GP, Sethi CS, Linberg KA, Charteris DG, Fisher SK: Experimental retinal reattachment: a new perspective. Mol Neurobiol 2003;28:159–175.

19 Jaynes CD, Sheedlo HJ, Agarwal N, O'Rourke K, Turner JE: Müller and retinal pigment epithelial cells (RPE) cell expression of NGFI-A and c-fos mRNA in response to medium conditioned by the RPE. Brain Res Mol Brain Res 1995;32:329–337.

20 Hollborn M, Tenckhoff S, Jahn K, Iandiev I, Biedermann B, Schnurrbusch UE, Limb GA, Reichenbach A, Wolf S, Wiedemann P, Kohen L, Bringmann A: Changes in retinal gene expression in proliferative vitreoretinopathy: glial cell expression of HB-EGF. Mol Vis 2005;11:397–413.

21 Lewis GP, Erickson PA, Guerin CJ, Anderson DH, Fisher SK. Basic fibroblast growth factor: a potential regulator of proliferation and intermediate filament expression in the retina. J Neurosci 1992;12:3968–3978.

22 Shen WG, Peng WX, Shao Y, Xu JF, Dai G, Zhang Y, Pan FY, Li CJ: Localization and activity of calmodulin is involved in cell-cell adhesion of tumor cells and endothelial cells in response to hypoxic stress. Cell Biol Toxicol 2007;23:323–335.

23 Lewis GP, Mervin K, Valter K, Maslim J, Kappel PJ, Stone J, Fisher S: Limiting the proliferation and reactivity of retinal Müller cells during experimental retinal detachment: the value of oxygen supplementation. Am J Ophthalmol 1999;128:165–172.

24 Kirchhof B: Strategies to influence PVR development. Graefes Arch Clin Exp Ophthalmol 2004;242: 699–703.

25 Cai J, Wei R, Ma X, Zhu H, Li Y: Cytotoxic effects of antiproliferative agents on human retinal glial cells in vitro. Int Ophthalmol 2001:24:225–231

26 Charteris DG, Aylward GW, Wong D, Groenewald C, Asaria RH, Bunce C, PVR Study Group: A randomised controlled trial of combined 5-fluouracil and low-molecular weight heparin in management of established proliferative vitreoretinopathy. Ophthalmology 2004;111:2240–2245.

27 Wickham L, Bunce C, Wong D, McGurn D, Charteris DG: Randomized controlled trial of combined 5-fluouracil and low-molecular weight heparin in the management of unselected rhegmatogenous retinal detachments undergoing primary vitrectomy. Ophthalmology 2007;114:698–704.

28 Sundar S, Jha TK, Thakur CP, Engel J, Sindermann H, Fischer C, Junge K, Bryceson A, Berman J: Oral miltefosine for Indian visceral leishmaniasis. N Engl J Med 2002;347:1739–1746.

29 Leonard R, Hardy J, van Tienhoven G, Houston S, Simmonds P, David M and Mansi J. Randomized, double-blind, placebo-controlled, multicenter trial of 6% miltefosine solution, a topical chemotherapy in cutaneous metastases from breast cancer. J Clin Oncol. 2001;21:4150–4159.

30 Eibl KH, Lewis GP, Betts K, Linberg KA, Gandorfer A, Kampik A, Fisher SK: The effect of alkylphosphocholines on intraretinal proliferation initiated by experimental retinal detachment. Invest Ophthalmol Vis Sci 2007;48:1305–1311.

31 Eibl KH, Banas B, Kook D, Ohlmann AV, Priglinger S, Kampik A, Welge-Luessen UC: Alkylphosphocholines: a new therapeutic option in glaucoma filtration surgery. Invest Ophthalmol Vis Sci 2004;45:2619–2624.

32 Eibl KH, Banas B, Schoenfeld CL, Neubauer AS, Priglinger S, Kampik A, Welge-Luessen U: Alkylphosphocholines inhibit proliferation of human retinal pigment epithelium. Invest Ophthalmol Vis Sci 2003;44:3556–3561.

33 Shoji M, Raynor RL, Fleer EA, Eibl H, Vogler WR, Kuo JF: Effects of hexadecylphosphocholine on protein kinase C and TPA-induced differentiation of HL60 cells. Lipids 1991;26:145–149.

34 Wu TT, Hsieh YH, Hsieh YS, Liu JY: Reduction of PKCalpha decreases cell proliferation, migration, and invasion of human malignant hepatocellular carcinoma. J Cell Biochem 2008;103:9–20.

35 Lipinski CA, Tran NL, Bay C, Kloss J, McDonough WS, Beaudry C, Berens ME, Loftus JC: Differential role of praline-rich tyrosine kinase 2 and focal adhesion kinase in determining glioblastoma migration and proliferation. Mol Cancer Res 2003;1:323–332.

36 Shen TL, Guan JL: Differential regulation of cell migration and cell cycle progression by FAK complexes with Src, PI3K, Grb7 and Grb2 in focal contacts. FEBS Lett 2001;499:176–181.

37 Ruiter GA, Zerp SF, Bartelink H, van Blitterswijk WJ, Verheij M: Anti-cancer alkyl-lysophospholipids inhibit the phosphatidylinositol 3-kinase-Akt/PKB survival pathway. Anticancer Drugs 2003;14:167–173.

38 Ozaki M, Ogita H, Takai Y: Involvement of integrin-induced activation of protein kinase C in the formation of adherens junctions. Genes Cells 2007;12: 651–662.

39 Eibl KH, Kook D, Priglinger S, Haritoglou C, Kampik A, Welge-Luessen U: Human retinal pigment epithelial cell attachment, spreading and migration inhibition by alkylphosphocholines. Invest Ophthalmol Vis Sci 2006, 47:364–370.

40 Schuettauf F, Eibl KH, Thaler S, Shinoda K, Rejdak R, May CA, Blatsios G, Welge-Luessen U: Toxicity study of erucylphosphocholine in a rat model. Curr Eye Res 2005;30:813–820.

41 Eibl KH, Schwabe K, Welge-Luessen U, Kampik A, Eichler W. The role of alkylphosphocholines in retinal Müller glial cell proliferation. Curr Eye Res 2008;33:385–393.

Kirsten H. Eibl, MD, FEBO
University Eye Hospital, Ludwig-Maximilians-University
Mathildenstrasse 8
DE–80336 Munich (Germany)
Tel. +49 89 5160 3800, Fax +49 89 5160 4788, E-Mail Kirsten.Eibl@med.uni-muenchen.de

Gandorfer A (ed): Pharmacology and Vitreoretinal Surgery.
Dev Ophthalmol. Basel, Karger, 2009, vol 44, pp 56–68

Neuroprotection for Diabetic Retinopathy

Hisanori Imai · Ravi S.J. Singh · Patrice E. Fort · Thomas W. Gardner

Penn State Hershey Eye Center and Department of Cellular and Molecular Physiology, Penn State College of Medicine, Hershey, Pa., USA

Abstract

Diabetic retinopathy (DR) is a neurodegenerative and microvascular disease resulting in functional and structural changes of all cell types in the retina. Several mechanisms for neuroretinal homeostasis, including the blood-retinal barrier, normal metabolite delivery into the retina, and the effect of neurotrophins for the retina, are impaired in DR. However, it is still not clear which components are most important for the development of DR and which may be most useful as therapeutic targets. In this chapter, we summarize the evidence for the neurodegeneration in DR and review normal mechanisms for maintenance of postmitotic cells in the retina and alterations in normal maintenance pathways in DR with emphasis on 'neuroprotection'. Finally, we discuss current neuroprotective strategies and future directions for the treatment of DR.

Until recently, the concept of 'neuroprotection' for diabetic retinopathy (DR) would have seemed implausible given that the view of the disease as a purely microvascular problem prevailed. We currently consider DR as a 'neurovascular' disorder [1] since the retina is part of the nervous system, because neurons are the cells that subserve visual function and must by definition be altered when vision is lost, as in DR.

In this review, we emphasize an integrated view of the retina (a 'network') in which the blood-retinal barrier (BRB) is vital for neural retinal function and vice versa. Therefore, maintenance of microvascular integrity may be considered to be neuroprotective as much as strategies that target neurons or glial cells directly. Likewise, regulation of normal anabolic and synthetic processes, such as glucose metabolism, and lipid, nuclei acid and protein synthesis is 'neurotrophic' as much as the actions of peptide growth factors. Therefore, this review defines neuroprotection broadly and with the admission that we do not fully understand the mechanisms that maintain the retina in its highly differentiated state so it can perform its task of vision.

Evidence for Neurodegeneration in Diabetic Retinopathy

DR is a common complication of diabetes and approximately 75% of persons with type 1 diabetes develop retinopathy, while approximately 50% of persons with type 2 diabetes may develop retinopathy [2], and it is the leading cause of new cases of legal blindness in Americans between the ages of 20 and 74 years [3]. Neurodegeneration is recognized as a component of some retinal diseases including glaucoma, retinitis pigmentosa [4], and while there has been an emphasis on changes to the retinal vasculature in diabetes, recent findings reveal neurodegeneration of the retina in DR.

Retinal function changes before the onset of clinically detectable vascular lesions associated with DR. For example, the amplitude and latency of oscillatory potentials in the electroretinogram is impaired in both humans and animals with diabetes [5], and the reasons for these changes have been obscure, but recent cell biological studies suggest possible explanations. For example, the number of apoptotic nuclei in the retina is elevated after only 4 weeks of streptozotocin-induced diabetes in the rat [6], the rate of apoptosis is essentially constant over 12 months of diabetes, and the amount of apoptosis was reduced by the treatment with subcutaneous insulin implants. The majority of TUNEL-positive cells did not colocalize with an endothelial cell-specific marker. This finding suggests that most of the cells undergoing apoptosis were not vascular and more likely to be either neurons or glial cells. Consistent with this finding, the number of retinal ganglion cells (RGCs) and the thickness of the inner retina were reduced after 8 months of diabetes, and the total number of RGC bodies was reduced by 10% compared to age-matched controls, suggesting a reduction in the total number of neurons. There was also a 50% reduction in the total number of cells exhibiting immunoreactivity for NeuN, a cell-specific marker expressed in the nuclei of neurons, in rats with 4 months' diabetes, and a 20% reduction in the NeuN-positive RGCs. The loss of RGCs is also reflected by a reduction in the number of axons in the optic nerve. A similar increase in apoptosis was also observed in postmortem retinas from patients with diabetes. At scanning laser polarimetry, the thickness of the nerve fiber layer in the superior polar quadrant of the retina was significantly reduced in the group of patients with diabetes compared to the control group [7]. Thus, these results clearly indicate that diabetes induces a chronic neurodegeneration in the retina, and although the relationship to specific visual defects remains uncertain, it is likely they contribute directly to altered function in some way.

Normal Mechanisms for Maintenance of Postmitotic Neurons in the Retina

Intact Blood-Retinal Barrier

The BRB is required to maintain precise neuronal and glial cell exchange of neurotransmitters and metabolites in the neural retina. The BRB is formed by the retinal pigmented epithelium (RPE) and the blood vessels of the retina, which form the outer and inner barriers, respectively. The RPE controls the flow of solutes and fluid from the choroidal vasculature into the outer retina and the retinal vasculature directly controls flux into the inner retina. This strict control of solutes and fluids that cross the BRB is achieved through well-developed intercellular tight junctions. Tight junctions control flux across both endothelial and epithelial cells creating defined environments such as the neural retina. They provide a barrier function restricting the paracellular flux of molecules and a fence function maintaining apical and basolateral membrane composition within a cell. Over 40 proteins are associated with tight junctions, including occludin, claudins and zonula occludens. Occludin and the family of claudins, both of which are transmembrane proteins, are responsible for the direct cell-to-cell attachment in the establishment of the tight junction barrier. Occludin is expressed predominantly on epithelial and endothelial cells and its content correlates with the degree of barrier properties [8]. As many as 23 isoforms of claudins have been described and shown to be implicated in the control of ion and small molecule movement through the tight junction barrier [9]. The wide distribution of different claudins in various tissues with a wide range of different barrier properties suggests that these proteins control much of the tissue barriers characteristics. Zonula occludens are intracellular proteins that associate with the cytoplasmic surface of the tight junctions and organize the tight junction complex. Thus, the BRB contributes to the normal retinal homeostasis and tight junction proteins are important dynamic regulators of the BRB.

Regulated Metabolite Delivery

Glucose is an essential metabolic substrate of all mammalian cells and neural tissue including the retina depends entirely on glucose under normal physiological conditions. The glucose supply to the retina derives entirely from the circulation, and both outer and inner BRBs regulate the passive diffusion of glucose and other bloodborne nutrients into the interstitial space surrounding neural cells. Glucose entry is mediated by specific transport process involving members of the sodium-independent glucose transporter (GLUT) family. Five functional isoforms of the GLUT family have been identified. In the retina, GLUT1 is expressed in the retinal capillary endothelial cells and the RPE. In the human inner BRB, a substantial portion of total cellular GLUT1 resides in cytosolic stores. These transporters may rapidly translocate to the plasma membrane where they mediate glucose entry into endothelial cells in response to external stimulation. GLUT1 is also expressed in RGCs, photoreceptors and Müller cells. The finding of abundant expression of GLUT1 in Müller cells is in line with evidence of glucose transport and metabolism in isolated cell preparations. Glial cells support neuron function in numerous ways, so this result illustrates one way in which Müller cells serves as a nutrient-supporting cell of the retina [10].

In addition to controlling retinal glucose delivery, the BRB also regulates access of amino acids to the retina that are used as neurotransmitters and protein precursors. The BRB prevents excessive retinal access by circulating antibodies and immune cells.

Excitatory Neurotransmittors

Amino acids or their metabolites have a role as neurotransmitters and are involved in normal synaptic functions under physiologic conditions. Glutamate works as an excitatory neurotransmitter in the central nervous system, including the retina [11], in which photoreceptors, bipolar cells and ganglion cells release glutamate. The synaptic function by glutamate is mediated through various postsynaptic receptors, including the N-methl- D-aspartate (NMDA) receptor, and terminated by its re-uptake from the synaptic cleft into surrounding cells by the excitatory amino acid transporter (EAAT). Meanwhile, if excessive amounts of glutamate are released from injured nerve cells or if glutamate re-uptake by EAAT is not sufficient, neuronal death can result in a process known as glutamate-mediated excitotoxicity. The NMDA receptors are associated with this excitotoxicity and calcium entry into the cells, which finally causes apoptosis of the cells, and the re-uptake system by EAAT is essential to keep the extracellular glutamate concentration, to control the specificity of glutamate effects and to limit the potentially damaging effects of lingering glutamate, including the neurodegeneration [11]. Thus, in retina, glutamate contributes to both the physiological synaptic function and excitotoxicity induced in several ocular neurodegenerative disease including glaucoma and DR.

Neurotrophins

Vascular Endothelial Growth Factor

Vascular endothelial growth factor (VEGF) is protein identified initially as an endothelial cell mitogen and vascular permeability factor. In the retina, VEGF is produced by the RPE, pericytes, RGCs, Müller cells and endothelial cells in normal conditions and its expression is regulated in response to hypoxic conditions. It is well known that VEGF is a mediator of angiogenesis, BRB breakdown, inflammation, and leukocyte adhesion. In addition, VEGF affects neuronal growth, differentiation and survival. In vitro, VEGF stimulates axonal outgrowth, improves the survival of ganglion neurons, and can rescue hippocampal neurons from apoptosis induced by serum withdrawal. In vivo, VEGF coordinates migration of motor neuron soma and local delivery of VEGF prolongs motor neuron survival. Conversely, low VEGF levels have been linked to the neuronal degeneration in both animal models and human disease [12]. These results suggest an important role for VEGF in neuronal development and maintenance within the central nervous system. The receptors for VEGF are present in normal retinal neuronal cells, indicating a possible functional role for VEGF in the neural retina [13]. Gene expression studies in the brain and retina also suggest that VEGF is upregulated by hypoxia preconditioning, a brief ischemic episode that protects neurons, against subsequent prolonged ischemia-reperfusion-related damage. Thus, VEGF might provide neuroprotection in the retina, particularly during ischemia or other metabolic stresses. In fact, glucosamine stimulates VEGF expression in retinal neurons, possibly via RNA-dependent protein kinase-like endoplasmic-reticulum associated kinase (PERK) [14], so ischemia is not the only stimulus for VEGF expression.

Erythropoietin

Erythropoietin (EPO) is a member of the cytokine superfamily, and has been known best as a hematopoietic cytokine produced by fetal liver and adult kidney. Recently, it has been shown that EPO and EPO receptor are also expressed in the human central nervous system including retina. In a model of retinal ischemia, EPO prevented apoptotic cell degeneration, and improved functional recovery. Likewise, light-induced photoreceptor degeneration was reduced by systemic pre- or post-treatment with recombinant EPO [15]. EPO-mediated neurotrophic effects have been attributed to pleiotropic mechanisms, including interactions with Jak2, Akt/PKB, ERK1/2, and Bcl-XL, increased calcium influx, reduction of nitric oxide (NO) production, restoration of glutathione peroxidase activity, and IL-6 suppression.

Insulin and Insulin-Like Growth Factors

Insulin is a potent anabolic hormone synthesized in the pancreatic beta cells that stimulates the uptake and storage of carbohydrates, fatty acids, and amino acids into glycogen, fat, and protein, respectively. The insulin supply to retina is thought to depend on the delivery from the circulation, although several studies have suggested the possibility of insulin production in other extrapancreatic tissues including retinas [16]. However, our studies have not found detectable levels of pre-proinsulin mRNA in normal rat retina [unpubl. data]. In normal conditions, insulin-like growth factor-1 (IGF1) and IGF2, which are 60% homologous at the amino acid level and have significant homology to insulin, are expressed in several tissues including the retina. IGF1 can mimic the metabolic effect of insulin. The activity of all three ligands is mediated through the insulin receptor (IR) and insulin-like growth factor receptor-1 (IGF1R), both belonging to the transmembrane tyrosine kinase receptor family. Ligand binding and subsequent autophosphorylation of IR and/or IGF1R leads to activation of several intracellular signaling pathways including phosphoinositide-3 kinase (PI3-K) and/or mitogen-activated protein kinase (MAPK).

We demonstrated Akt-dependent cell survival in cultured retinal neurons in response to insulin and IGF1 [17] in a PI3-K/Akt–dependent manner and reduced apoptosis induced by serum deprivation. Physiological concentration of insulin increases IR beta autophosphorylation, Akt^{Ser473} phosphorylation in the retinal nuclear layers and Akt-1 kinase activity in ex vivo cultures [18]. Inhibition of p70 ribosomal protein S6 kinase (p70S6K) activity, a protein synthesis regulator, using a dominant negative mutant, prevented insulin-induced cell survival, whereas overexpression of wild-type p70S6K enhanced the effect of insulin on survival. Meanwhile, IGF1 and its analog inhibited the neuroretinal cell death in vitro, and intraocular injection of IGF1 protected retinal ganglion cells from axotomized-induced secondary death via PI3-K dependent Akt phosphorylation and by inhibition of caspase-3

[19]. Thus, insulin and IGF1 have a neuroprotective effects via the IR/PI3-K pathway but the physiologic roles of these ligands remains incompletely defined.

Pigment Epithelium-Derived Factor

Pigment epithelium-derived factor (PEDF) is a 50-kDa glycoprotein that belongs to the superfamily of serine protease inhibitors. PEDF was first identified in conditioned medium from fetal human RPE cell cultures with neuronal differentiating activity [20], and is widely synthesized in neural and non-neural tissues. In the retina, its expression is detected in RPE, photoreceptor and ganglion cells. PEDF can promote activation of nuclear factor κB (NF-κB), a transcription factor that in turn induces the expression of antiapoptotic genes and/or neurotrophic factors in cerebellar granule cells. PEDF triggers NF-κB activation through increased phosphorylation and decreased protein levels of the inhibitor of NF-κB (IκB) as part of its mechanism to protect immature cerebellar granule cells against apoptosis. PEDF also protects rat retinal neurons from hydrogen peroxide-induced cell death in culture. PEDF delays the death of photoreceptor cells in models of retinitis pigmentosa, retinal degeneration (rd/rd), and retinal degeneration slow (rds/rds) mice [21]. PEDF protects rat photoreceptors cells from light damage [22]. PEDF also protects retinal neurons from transient ischemic reperfusion [23]. Thus, PEDF has neuroprotective roles in both acute and chronic forms of neurodegeneration.

Brain-Derived Neurotrophic Factor

Brain-derived neurotrophic factor (BDNF) is widely expressed throughout the central and peripheral nervous systems. In the retina, BDNF is expressed in RGCs, amacrine and Müller cells [24]. BDNF binds with high affinity to the TrkB receptor, which is expressed in RGCs, amacrine cells and the RPE, and in cones but not rods. TrkB receptor signaling by BDNF occurs through both PI3-K and MEK/ERK pathways, and suppresses the activation of caspase-3. BDNF provides autocrine and paracrine trophic support to RGCs. Upregulation of Trk receptor gene expression appears to be a normal response of RGCs to axotomy and BDNF enhances the survival of RGCs after optic nerve injury [25]. Transgenic expression of BDNF prolongs the survival of RGCs after experimental axotomy [25], and axon-mediated gene transfer of a Trk receptor into RGCs via retrograde transport from axonal terminal arbors delays axotomy-induced apoptosis [26]. TrkB gene transfer into RGCs together with administration of exogenous BDNF also increases RGCs survival. Thus, BDNF is involved in normal maintenance of the retina.

Alterations in Normal Cellular Maintenance Pathways in Diabetic Retinopathy

Dysregulated Blood-Retinal Barrier

Breakdown of the BRB occurs early in the course of diabetes, before retinopathy is observed clinically. Several permeabilizing factors such as histamine, IL-1, tumor necrosis factor-α and VEGF have been implicated as causes of breakdown of the BRB [27]. The reduced expression and anatomic distribution of occludin in experimental diabetes occur in a timeframe concurrent with early neurodegenerative changes. A similar decrease in occludin and ZO-1 tight junction protein content was also noted upon VEGF treatment of cultured brain microvessel endothelial cells, coincident with increase permeability of the cellular monolayer. Redistribution of occludin from the plasma membrane to the endothelial cell cytoplasm, a phenomenon that is accompanied by increased occludin phosphorylation, is observed with both diabetes and with VEGF treatment [28]. Intravitreal injection of VEGF in rats increased phosphorylation of ZO-1 and occludin. All of these modifications may be critical for the breakdown of the BRB in diabetes, and VEGF likely facilitates this breakdown.

Altered Glucose Metabolism

Glucose accumulates to a supranormal intracellular level in diabetes as a result of an increased uptake, a decreased efflux, or metabolism of intracellular glucose. Retinal glucose also increases in diabetes in the extracellular space as a result of increased permeability of the BRB [29]. GLUT1 mRNA abundance has been reported to be increased in the retina of diabetic rats. In streptozotocin-induced diabetic rats, decreased total retinal GLUT1 and retinal microvascular GLUT1 were reported [30]. On the other hand, a localized GLUT1 increase on endothelial cells of the inner BRB was demonstrated in postmortem retina from patients with diabetes [31]. These results may suggest that the upregulation of particular inner BRB GLUT1 occurs in the early stages of DR, and that the resultant increase in inner BRB glucose transport could play a role in the progression of the DR.

Altered Glutamate Metabolism

As mentioned above, glutamate-mediated excitotoxicity has been shown to play an important role in several ocular diseases including DR. Elevated glutamate concentration was detected in human vitreous samples in patients with PDR and in the retina of streptozotocin-induced diabetic rats. At the same time, reduced glutamate oxidation and glutamine synthesis is due to transaminase and glutamine synthetase dysfunction in streptozotocin-induced diabetic rats [32]. Additionally, increased immunoreactivity of NMDA receptor subunit-1 was observed in both outer and inner plexiform layers in donor eyes from subjects with diabetes [33], and in ganglion, amacrine and bipolar cells in streptozotocin-induced diabetic rats. Moreover, significantly decreased activity of the Müller cell glutamate transporter, as analyzed by the patch-clamp technique, was detected in streptozotocin-induced diabetic rats. Taken together, these data suggest that the abnormal glutamate metabolism induced by diabetes may contribute to the neurodegeneration in DR.

Disturbances of Neurotrophin Function

As detailed in the preceding paragraphs, peptide growth factors are expressed throughout the retina as part of normal homeostasis. So what are the implications of increased expression of these factors in DR? We hypothesize [34] that they increase as an adaptive response to the metabolic alterations of diabetes in an attempt to maintain retinal viability, but a maladaptive state ensues when the diabetes persists, resulting in excessive growth factor production. This overproduction may then result in the clinical features of macular edema and neovascularization. Moreover, merely finding an increased ligand concentration does not necessarily imply increased ligand receptor activity because soluble receptors can also be overexpressed or the receptor activity may be downregulated [35]. Thus, neurotrophins serve multiple roles and their net effects may depend on the context of other cellular events.

VEGF

Clinical studies have revealed that intravitreal VEGF concentration increases in subjects as they progress from nonproliferative DR to active proliferative diabetic retinopathy (PDR) [36]. In diabetic rats, the levels of VEGF and its receptor VEGFR2 in the retina are increased [13]. The increased VEGF expression is strongly correlated with increased vascular permeability following the breakdown of BRB and the neovascularization.

Erythropoietin

An elevated EPO concentration was detected in human vitreous samples in patients with PDR and diabetic macular edema [37]. Additionally, the expression of the EPO receptor was increased in streptozotocin-induced diabetic rat retinas, whereas the expression of the EPO protein was unchanged. These results suggest that the EPO and EPO receptor which are induced in hypoxic conditions are also induced during the early stages of diabetes.

Insulin/Insulin-Like Growth Factors

The intravitreal insulin concentration was unchanged in patients with type 2 diabetes who were not receiving insulin treatment compared to controls [38]. On the other hand, a reduced intravitreal insulin concentration was detected in streptozotocin-induced diabetic rats. These results may suggest that type 2 diabetic patients may have normal vitreous insulin concentration, whereas type 1 patients may have reduced vitreous insulin, but the vitreous insulin levels were not corrected for plasma insulin concentrations in the human study. Studies of IGF1 concentration in the serum in diabetic subjects are still controversial. Several studies suggest that patients with PDR have elevated levels of IGF in serum [39]. On the other hand, patients with type 1 diabetes usually demonstrate reduced levels of serum IGF1 because of decreased hepatic production [40]. The existence of insulin-like growth factor-binding proteins makes it difficult to measure the correct concentration of free IGFs and could explain these discrepancies. In the vitreous, an increased concentration of IGF1 was detected in patients with PDR [41].

The abundance of IGF1 mRNA was significantly higher in retina from diabetic rats compared to age-matched controls. On the other hand, reduced retinal levels of IGF1 protein were detected in diabetic animals. These differences may be due to differences in the duration of diabetes as well as the type of diabetes in the animals.

Pigment Epithelium-Derived Factor

Intravitreal PEDF concentration was decreased in patients with PDR and in experimental diabetic rats [42], whereas it was significantly higher in experimental diabetic rat retinal tissue. On the other hand, inhibition of VEGF-induced and advanced glycation end products (AGE)-elicited retinal leukostasis by PEDF was reported. This result suggests that controlling PEDF levels might be a valuable approach to managing some neurological diseases including DR.

Brain-Derived Neurotrophic Factor

In diabetic rat retina, BDNF protein expression was decreased whereas TrkB receptor expression was not altered [24]. BDNF immunoreactivity was decreased in diabetic rat retinas as were TH-1 and Thy-1 protein levels, markers of dopaminergic amacrine cells and RGC cells, respectively. Intravitreal BDNF injection restored TH-1 protein expression and the TH-1-positive cell density. These results suggest the involvement of a reduced BDNF expression in the development of DR and specifically its neurodegenerative aspect.

Neuroprotective Strategies for Diabetic Retinopathy Treatment

Control of Retinal Glucose Metabolism

Glucose is the primary energy substrate in the retina so appropriate regulation of its metabolism is central to the viability of retinal cells. Impaired glucose homeostasis is a major systemic consequence of the altered insulin action in diabetes. Retinal glucose concentrations are increased in diabetic rats [43] but glycolytic flux is reduced in retinas of diabetic animals compared to controls [44]. Moreover, retinal glucose metabolism is linked closely with that of glutamate and branched chain amino acids [45] and thus very complex. However, at this point there are insufficient data to provide a thorough understanding of the defects in retinal glucose metabolism in either type 1 or type 2 diabetes.

Actions of Systemically Administered Insulin

Diabetic retinopathy in experimental rats involves loss of constitutive retinal insulin receptor signaling that is restored by systemic and intravitreal insulin [18]. Insulin provides trophic support for retinal neurons in culture [17] and systemic insulin treatment reduces retinal apoptosis in diabetic rats [6]. Intrathecal delivery of low-dose insulin, insufficient to reduce hyperglycemia, improves peripheral nerve conduction velocity in diabetic rats and sequestering intrathecal insulin in nondiabetic rats using an anti-insulin antibody slows conduction velocity and axonal atrophy resembling changes in diabetes. Thus, insulin appears to have direct neuroprotective effects for the central nervous system independent of those related to glucose metabolism.

The DCCT showed a lower risk of retinopathy progression in type 1 diabetic patients who received intensive diabetes therapy (3–4 insulin injections daily or a pump) than those receiving conventional treatment (1 or 2 daily injections) [46]. This reduced risk was achieved by administering insulin more frequently to the intensive treatment group and this lower risk persisted in the intensive treatment group patients for the same HbA1c levels as those in the conventional treatment group. Moreover, the risk of progression was lower in patients in whom insulin treatment was initiated at an earlier stage of disease, when no clinically evident retinopathy was present, compared to those in whom intensive treatment began after the onset of mild-to-moderate nonproliferative retinopathy. Hence, the DCCT might be viewed as an insulin dose-response trial [47]. Further clinical evidence for a role of insulin derives from the observation that systemic insulin resistance is a risk factor for retinopathy in patients with type 1 diabetes in the DCCT [48], the EURODIAB study [49], and the Pittsburgh Epidemiology of Diabetes Complications study [50]. In type 2 diabetes, insulin deficiency is also an independent risk factor for the presence of retinopathy [51]. These data suggest that the retina is an insulin sensitive tissue and retinal insulin signaling provides neurotrophic support.

Sigma Receptor 1 and Its Ligand, (+)-Pentazocine

Sustained (+)-pentazocine treatment in $Ins2^{Akita/+}$ diabetic mice restored the disruptions of retinal architectures induced by DR, like decrease of the inner plexiform layer and inner nuclear layer thickness, reduction of ganglion cells, and loss of the uniform organization of radial Müller fibers [52]. This result may suggest the potential use of (+)-pentazocine, which is a ligand of sigma receptor 1, for the treatment of neurodegenerative disease including DR.

Future Directions

Phase 1 trials of ciliary neurotrophic factor (CNTF) for treatment of retinitis pigmentosa highlight the promising potential of neurorophins to treat retinal neurodegeneration [53]. Of all the major retinal diseases, DR is the only one for which large-scale clinical trials have proven the ability to reduce disease incidence and progression [46]. Neurodegeneration in DR starts before the onset of clinically evident disease [6] and this early phase of disease offers the greatest window of opportunity for treatment before significant vision loss has occurred. Any therapeutic neurotrophin must act to restore retinal physiology without serious side effects, and it is plausible that neurotrophins may have different therapeutic potentials at different stages of DR. It is unlikely that a single neurotrophic agent would correct the myriad of neurovascular degenerative changes in DR. Further, neurotrophin changes in DR such as increased VEGF medicated vascular permeability may represent adaptive responses that have become maladaptive over time. A better understanding of the role of neurotrophins in retinal physiology would be key to devising therapeutic interventions in disease.

Acknowledgement

Juvenile Diabetes Research Foundation, American Diabetic Association, Pennsylvania Lions Sight Conservation and Eye research Foundation. Thomas W. Gardner is the 'Jack and Nancy Turner Professor'.

References

1 Gardner TW, Antonetti DA, Barber AJ, LaNoue KF, Levison SW: Diabetic retinopathy: more than meets the eye. Surv Ophthalmol 2002;47(suppl 2):S253–S262.

2 Kempen JH, O'Colmain BJ, Leske MC, Haffner SM, Klein R, Moss SE, Taylor HR, Hamman RF: The prevalence of diabetic retinopathy among adults in the United States. Arch Ophthalmol 2004;122:552–563.

3 Klein R, Klein B: Vision disorders in diabetes. National Diabetes Data Group Diabetes in America. NIH Bethesda, NIH Publication No 95–1468. 1995, pp 293–337.

4 Baumgartner WA: Etiology, pathogenesis, and experimental treatment of retinitis pigmentosa. Med Hypotheses 2000;54:814–824.

5 Juen S, Kieselbach GF: Electrophysiological changes in juvenile diabetics without retinopathy. Arch Ophthalmol 1990;108:372–375.

6 Barber AJ, Lieth E, Khin SA, Antonetti DA, Buchanan AG, Gardner TW: Neural apoptosis in the retina during experimental and human diabetes: early onset and effect of insulin. J Clin Invest 199815;102:783–791.

7 Lopes de Faria JM, Russ H, Costa VP: Retinal nerve fibre layer loss in patients with type 1 diabetes mellitus without retinopathy. Br J Ophthalmol 2002;86: 725–728.

8 Mitic LL, Anderson JM: Molecular architecture of tight junctions. Annu Rev Physiol 1998;60:121–142.

9 Turksen K, Troy TC: Barriers built on claudins. J Cell Sci 2004;117:2435–2447.

10 Tsacopoulos M, Poitry-Yamate CL, Poitry S, Perrottet P, Veuthey AL: The nutritive function of glia is regulated by signals released by neurons. Glia 1997;21:84–91.

11 Barnstable CJ: Glutamate and GABA in retinal circuitry. Curr Opin Neurobiol 1993;3:520–525.

12 Lambrechts D, Storkebaum E, Morimoto M, Del-Favero J, Desmet F, Marklund SL, Wyns S, Thijs V, Andersson J, van Marion I, Al-Chalabi A, Bornes S, Musson R, Hansen V, Beckman L, Adolfsson R, Pall HS, Prats H, Vermeire S, Rutgeerts P, Katayama S, Awata T, Leigh N, Lang-Lazdunski L, Dewerchin M, Shaw C, Moons L, Vlietinck R, Morrison KE, Robberecht W, Van Broeckhoven C, Collen D, Andersen PM, Carmeliet P: VEGF is a modifier of amyotrophic lateral sclerosis in mice and humans and protects motoneurons against ischemic death. Nat Genet 2003;34:383–394.

13 Gilbert RE, Vranes D, Berka JL, Kelly DJ, Cox A, Wu LL, Stacker SA, Cooper ME: Vascular endothelial growth factor and its receptors in control and diabetic rat eyes. Lab Invest 1998;78:1017–1027.

14 Kline CL, Schrufer TL, Jefferson LS, Kimball SR: Glucosamine-induced phosphorylation of the alpha-subunit of eukaryotic initiation factor 2 is mediated by the protein kinase R-like endoplasmic-reticulum associated kinase. Int J Biochem Cell Biol 2006;38:1004–1014.

15 Grimm C, Wenzel A, Groszer M, Mayser H, Seeliger M, Samardzija M, Bauer C, Gassmann M, Reme CE: HIF-1-induced erythropoietin in the hypoxic retina protects against light-induced retinal degeneration. Nat Med 2002;8:718–724.

16 Budd GC, Pansky B, Glatzer L: Preproinsulin mRNA in the rat eye. Invest Ophthalmol Vis Sci 1993;34: 463–469.

17 Barber AJ, Nakamura M, Wolpert EB, Reiter CE, Seigel GM, Antonetti DA, Gardner TW: Insulin rescues retinal neurons from apoptosis by a phosphatidylinositol 3-kinase/Akt-mediated mechanism that reduces the activation of caspase-3. J Biol Chem 2001;276:32814–32821.
18 Reiter CE, Sandirasegarane L, Wolpert EB, Klinger M, Simpson IA, Barber AJ, Antonetti DA, Kester M, Gardner TW: Characterization of insulin signaling in rat retina in vivo and ex vivo. Am J Physiol Endocrinol Metab 2003;285:E763–774.
19 Seigel GM, Chiu L, Paxhia A: Inhibition of neuroretinal cell death by insulin-like growth factor-1 and its analogs. Mol Vis 2000;6:157–163.
20 Tombran-Tink J, Johnson LV: Neuronal differentiation of retinoblastoma cells induced by medium conditioned by human RPE cells. Invest Ophthalmol Vis Sci. 1989;30:1700–1707.
21 Cayouette M, Smith SB, Becerra SP, Gravel C: Pigment epithelium-derived factor delays the death of photoreceptors in mouse models of inherited retinal degenerations. Neurobiol Dis 1999;6:523–532.
22 Cao W, Tombran-Tink J, Elias R, Sezate S, Mrazek D, McGinnis JF: In vivo protection of photoreceptors from light damage by pigment epithelium-derived factor. Invest Ophthalmol Vis Sci 2001;42:1646–1652.
23 Ogata N, Wang L, Jo N, Tombran-Tink J, Takahashi K, Mrazek D, Matsumura M: Pigment epithelium derived factor as a neuroprotective agent against ischemic retinal injury. Curr Eye Res 2001;22:245–252.
24 Seki M, Tanaka T, Nawa H, Usui T, Fukuchi T, Ikeda K, Abe H, Takei N: Involvement of brain-derived neurotrophic factor in early retinal neuropathy of streptozotocin-induced diabetes in rats: therapeutic potential of brain-derived neurotrophic factor for dopaminergic amacrine cells. Diabetes 2004;53:2412–2419.
25 Mansour-Robaey S, Clarke DB, Wang YC, Bray GM, Aguayo AJ: Effects of ocular injury and administration of brain-derived neurotrophic factor on survival and regrowth of axotomized retinal ganglion cells. Proc Natl Acad Sci USA 1994;91:1632–1636.
26 Garcia Valenzuela E, Sharma SC: Rescue of retinal ganglion cells from axotomy-induced apoptosis through TRK oncogene transfer. Neuroreport 1998;9:3165–3170.
27 Armstrong D, Ueda T, Aljada A, Browne R, Fukuda S, Spengler R, Chou R, Hartnett M, Buch P, Dandona P, Sasisekharan R, Dorey CK: Lipid hydroperoxide stimulates retinal neovascularization in rabbit retina through expression of tumor necrosis factor-alpha, vascular endothelial growth factor and platelet-derived growth factor. Angiogenesis 1998;2:93–104.
28 Barber AJ, Antonetti DA, Gardner TW: Altered expression of retinal occludin and glial fibrillary acidic protein in experimental diabetes. The Penn State Retina Research Group. Invest Ophthalmol Vis Sci 2000;41:3561–3568.
29 Tang J, Zhu XW, Lust WD, Kern TS: Retina accumulates more glucose than does the embryologically similar cerebral cortex in diabetic rats. Diabetologia 2000;43:1417–1423.
30 Badr GA, Tang J, Ismail-Beigi F, Kern TS: Diabetes downregulates GLUT1 expression in the retina and its microvessels but not in the cerebral cortex or its microvessels. Diabetes 2000;49:1016–1021.
31 Kumagai AK, Vinores SA, Pardridge WM. Pathological upregulation of inner blood-retinal barrier Glut1 glucose transporter expression in diabetes mellitus. Brain Res 1996;706:313–317.
32 Lieth E, Barber AJ, Xu B, Dice C, Ratz MJ, Tanase D, Strother JM: Glial reactivity and impaired glutamate metabolism in short-term experimental diabetic retinopathy. Penn State Retina Research Group. Diabetes 1998;47:815–820.
33 Santiago AR, Hughes JM, Kamphuis W, Schlingemann RO, Ambrosio AF: Diabetes changes ionotropic glutamate receptor subunit expression level in the human retina. Brain Res 2008;1198:153–159.
34 Antonetti DA, Barber AJ, Bronson SK, Freeman WM, Gardner TW, Jefferson LS, Kester M, Kimball SR, Krady JK, LaNoue KF, Norbury CC, Quinn PG, Sandirasegarane L, Simpson IA: Diabetic retinopathy: seeing beyond glucose-induced microvascular disease. Diabetes 2006;55:2401–2411.
35 Gariano RF, Gardner TW: Retinal angiogenesis in development and disease. Nature 2005;438:960–966.
36 Miller JW, Adamis AP, Aiello LP: Vascular endothelial growth factor in ocular neovascularization and proliferative diabetic retinopathy. Diabetes Metab Rev 1997;13:37–50.
37 Hernandez C, Fonollosa A, Garcia-Ramirez M, Higuera M, Catalan R, Miralles A, Garcia-Arumi J, Simo R: Erythropoietin is expressed in the human retina and it is highly elevated in the vitreous fluid of patients with diabetic macular edema. Diabetes Care 2006;29:2028–2033.
38 Boulton M, Gregor Z, McLeod D, Charteris D, Jarvis-Evans J, Moriarty P, Khaliq A, Foreman D, Allamby D, Bardsley B: Intravitreal growth factors in proliferative diabetic retinopathy: correlation with neovascular activity and glycaemic management. Br J Ophthalmol 1997;81:228–233.
39 Hyer SL, Sharp PS, Brooks RA, Burrin JM, Kohner EM: A two-year follow-up study of serum insulin-like growth factor-I in diabetics with retinopathy. Metabolism 1989;38:586–589.

40 Krsek M, Skrha J, Sucharda P, Justova V, Lacinova Z: Changes in IGF-I levels and its binding proteins in diabetes mellitus and obesity. Cas Lek Cesk 2003; 142:216–219.
41 Meyer-Schwickerath R, Pfeiffer A, Blum WF, Freyberger H, Klein M, Losche C, Rollmann R, Schatz H: Vitreous levels of the insulin-like growth factors I and II, and the insulin-like growth factor binding proteins 2 and 3, increase in neovascular eye disease. Studies in nondiabetic and diabetic subjects. J Clin Invest 1993;92:2620–2625.
42 Ogata N, Nishikawa M, Nishimura T, Mitsuma Y, Matsumura M: Unbalanced vitreous levels of pigment epithelium-derived factor and vascular endothelial growth factor in diabetic retinopathy. Am J Ophthalmol 2002;134:348–353.
43 Tang J, Zhu XW, Lust WD, Kern TS: Retina accumulates more glucose than does the embryologically similar cerebral cortex in diabetic rats. Diabetologia 2000;43:1417–1423.
44 Ola MS, Berkich DA, Xu Y, King MT, Gardner TW, Simpson I, Lanoue KF: Analysis of glucose metabolism in diabetic rat retinas. Am J Physiol Endocrinol Metab 2006;290:E1057–1067.
45 Xu Y, Ola MS, Berkich DA, Gardner TW, Barber AJ, Palmieri F, Hutson SM, Lanoue KF: Energy sources for glutamate neurotransmission in the retina: absence of the aspartate/glutamate carrier produces reliance on glycolysis in glia. J Neurochem 2007;101: 120–131.
46 The Diabetes Control and Complications Trial: The effect of intensive diabetes treatment on the progression of diabetic retinopathy in insulin-dependent diabetes mellitus. Arch Ophthalmol 1995;113: 36–51.
47 Reiter CE, Gardner TW: Functions of insulin and insulin receptor signaling in retina: possible implications for diabetic retinopathy. Prog Retin Eye Res 2003;22:545–562.
48 Zhang L, Krzentowski G, Albert A, Lefebvre PJ: Risk of developing retinopathy in Diabetes Control and Complications Trial type 1 diabetic patients with good or poor metabolic control. Diabetes Care 2001;24:1275–1279.
49 Chaturvedi N, Sjoelie AK, Porta M, Aldington SJ, Fuller JH, Songini M, Kohner EM: Markers of insulin resistance are strong risk factors for retinopathy incidence in type 1 diabetes. Diabetes Care 2001;24: 284–289.
50 Stuhldreher WL, Becker DJ, Drash AL, Ellis D, Kuller LH, Wolfson SK, Orchard TJ: The association of waist/hip ratio with diabetes complications in an adult IDDM population. J Clin Epidemiol 1994; 47:447–456.
51 Kohner EM, Aldington SJ, Stratton IM, Manley SE, Holman RR, Matthews DR, Turner RC: United Kingdom Prospective Diabetes Study, 30: diabetic retinopathy at diagnosis of non-insulin-dependent diabetes mellitus and associated risk factors. Arch Ophthalmol 1998;116:297–303.
52 Smith SB, Duplantier J, Dun Y, Mysona B, Roon P, Martin PM, Ganapathy V: In vivo protection against retinal neurodegeneration by sigma receptor 1 ligand (+)-pentazocine. Invest Ophthalmol Vis Sci 2008;49:4154–4161.
53 Sieving PA, Caruso RC, Tao W, Coleman HR, Thompson DJ, Fullmer KR, Bush RA: Ciliary neurotrophic factor (CNTF) for human retinal degeneration: phase I trial of CNTF delivered by encapsulated cell intraocular implants. Proc Natl Acad Sci USA 2006;103:3896–3901.

Hisanori Imai
Penn State Hershey Eye Center and Department of Cellular and Molecular Physiology
Penn State College of Medicine
500 University Drive, HU097
Hershey, PA 17033 (USA)
Tel. +1 717 531 6711, Fax +1 717 531 0065, E-Mail himai@hmc.psu.edu

Gandorfer A (ed): Pharmacology and Vitreoretinal Surgery.
Dev Ophthalmol. Basel, Karger, 2009, vol 44, pp 69–81

VEGF Inhibitors and Vitrectomy for Diabetic Vitreoretinopathy

Carlos E. Cury, Jr[a] · Eduardo B. Rodrigues[a] · Carsten H. Meyer[b] · Michel E. Farah[a]

[a]Vision Institute, Department of Ophthalmology, Federal University of Sao Paulo, Sao Paulo, Brazil; [b]Department of Ophthalmology, University of Bonn, Bonn, Germany

Abstract

Diabetic retinopathy (DR) is one of the leading causes of blindness worldwide. Breakdown of the blood-retinal barrier in DR promotes accumulation of a high concentration of pre-retinal serum-derived chemoattractants, thereby stimulating cellular migration on the attached posterior hyaloid. This review assesses the role of intravitreal application of vascular endothelial growth factor (VEGF) inhibitors in combination with surgical removal of the vitreous. Vitrectomy with removal of the posterior hyaloids wash out the pocket of preretinal growth factors and enhance the diffusion of macromolecules, including VEGF, insulin-like growth factor 1 or histamines, from the retinal into the vitreous cavity for further absorption through the anterior segment outflow pathways. The release of tractional forces induced by the vitreomacular traction or epiretinal membranes demonstrated a strong correlation to the reduction of retinal thickness in DME. Three techniques are described to remove the pre-retinal thickened and adherent vitreous: (1) delamination, (2) segmentation, and (3) en block dissection. A better visualization of remaining cortex vitreous or adjacent epiretinal membrane and safer removal may be achieved by a better intraoperative visualization using a variety of vital dyes to satin retinal tissue (chromovitrectomy). Anti-VEGF treatments may represent an alternative adjunctive treatment for DR.

Diabetic retinopathy (DR) is one of the leading causes of blindness worldwide and the most common microvascular complication of diabetes [1]. During the first two decades of the disease, nearly all patients with type 1 diabetes and over 60% with type 2 diabetes develop retinopathy. In the Wisconsin Epidemiologic Study of Diabetic Retinopathy, also named WESDR, 3.6% of younger-onset patients and 1.6% of older-onset patients were blind [1]. Duration of diabetes and severity of hyperglycemia are the major risk factors for DR, while others include age, type of diabetes, clotting factors, and renal disease [2]. Timely application and reapplication of laser photocoagulation is the mainstay of treatment to reduce visual loss and to avoid the need for vitrectomy in patients with more advanced complications of diabetic retinopathy [3]. However, despite preventive regimens and timely treatment, substantial numbers of eyes will develop complications of proliferative retinopathy and may become candidates for vitrectomy [4].

Due to the limitations of current treatment approaches, new pharmacological therapies have being developed. The novel drugs target specific biochemical pathways that cause DR through involvement of protein kinase C (PKC) activation, oxidative stress, the angiogenesis pathway, and the glycation and sorbital pathway. These treatments aim to prevent diabetes-induced damage to the retinal microvasculature [5]. In this chapter the role of intravitreal application of vascular endothelial growth factor (VEGF) inhibitors in combination with surgical removal of the vitreous will be discussed.

Role of the Vitreous and Various Factors Including VEGF in Diabetic Vitreoretinopathy

In healthy eyes ultrastructural components of the vitreous can be observed histologically or histochemically, and basically they are mainly collagen type II, hyaluronan acid and hyalocytes [6]. In diabetic eyes, consecutive studies have demonstrated elevated levels of various cytokines such as VEGF, insulin-like growth factor-1 and histamines [7]. Pars plana vitrectomy with consecutive removal of the posterior hyaloids may resolve the diabetic macular edema (DME) and improve the central vision of patients who previously failed to respond to conventional laser photocoagulation treatment.

Four theories may support this hypothesis [7]:

Mechanical Release Theory. Breakdown of the BRB in DR promotes accumulation of a high concentration of preretinal serum-derived chemoattractants, thereby stimulating cellular migration on the attached posterior hyaloid. In addition, molecular changes in the vitreous induced by hyperglycemia may thicken and enhance its adherence to the retinal surface. Sebag et al. [9] suggested that the abnormal cross-linking might affect the collagen architecture structure in diabetic vitreous. The contraction of retinal cells, preretinal elements and the vitreous may induce tangential and anteroposterior traction and exacerbation of DME with vitreomacular traction (VMT) [8–10]. Release of tractional forces induced by the vitreous or epiretinal membranes demonstrated a strong correlation to the reduction of retinal thickness in DME. Moreover, several retrospective clinical studies have shown that vitrectomy and removal of the posterior hyaloid may lead to visual improvement of 2 lines in 38–92% of the eyes. Recently, Tachi and Ogino [11] observed the resolution of DME eyes within 3 months after vitrectomy without inner limiting membrane (ILM) peeling and questioned whether there is an additional beneficial effect of ILM removal.

Chemical Diffusion Theory. In eyes with diabetic retinopathy there may be an abnormal concentration of growth factors under the abnormally attached premacular posterior hyaloid thereby exacerbating DME. Vitrectomy with removal of the posterior hyaloids and vitreous gel may release and wash out the pocket of preretinal growth factors and enhance the diffusion of macromolecules, including VEGF, insulin-like growth factor 1 or histamines, from the retinal into the vitreous cavity for further absorption through the anterior segment outflow pathways. In 1992, Lewis et al. [12] described a positive effect on visual acuity of pars plana vitrectomy on diabetic patients who had a thickened and taut posterior hyaloid traction on the macula. In the following years, numerous retrospective studies have shown the advantage of vitrectomy and removal of the posterior hyaloid on morphologic and functional results [13, 14].

Decrease or Wash-Out of Pro-Angiogenic Growth Factors – VEGF. Previous reports by Funatsu's group have associated the aqueous VEGF concentration or determined its vitreous concentration in patients with DR and DME. A clear relationship has been established between structural and molecular parameters demonstrating that there are differences in the concentration of cytokines between DME and normal eyes. VEGF, a vascular endothelial cell mitogen and potent permeability factor, is produced by glial cells, retinal pigment epithelial cells, and vascular endothelial cells and is normally present in the retina and vitreous in low levels. Retinal hypoxia upregulates VEGF production, which results in abnormal angiogenesis, may increase in vascular permeability. Chronic hyperglycemia of uncontrolled diabetes leads to increased cellular levels of diacylglycerol, which in turn increases the synthesis of VEGF and also contributes to the microvascular abnormalities in DR. Inhibition of either VEGF moderates the microvascular complications seen in experimental animal models. Recent work also disclosed elevated levels of VEGF in ocular fluids of patients with proliferative diabetic retinopathy (PDR) [15]. These studies also point out the growth of new vessels from the retina or optic nerve to occur as a result of VEGF release into the vitreous cavity as a response to ischemia. Furthermore, injection of VEGF into normal primate eyes induces the same pathologic processes seen in diabetic retinopathy, including microaneurysm formation and increased vascular permeability [16]. In addition, experiments in animals have suggested a central role for the 165 isoform of VEGF specifically in the pathogenesis of DME, and increased retinal VEGF164 (the rodent equivalent to primate VEGF165) levels in this model coincide temporally with breakdown of the BRB [17]. Clinical studies demonstrate that intravitreal VEGF concentration increased as subjects progressed from early diabetic retinopathy to active PDR [16]. Gandorfer et al. [13] reported on the resolution of DME after vitrectomy in eyes without any evidence of VMT. The characterization of molecular and cellular events causing retinal microvascular abnormalities in DR has provided new targets for pharmacologic manipulation.

Oxygenation Hypothesis. Experimental data suggest that both scatter laser treatment and vitrectomy have a positive effect on the improvement of retinal oxygenation. Stefansson et al. [18], performed pars plana vitrectomy on one eye in cats and created a branch retinal vein occlusion in both eyes while the retinal oxygen tension was measured simultaneously in both. In the nonvitrectomized eye the branch retinal vein occlusion led to severe hypoxia of the retina. In the eye where vitrectomy had been performed there was insignificant change of the retinal oxygen tension with the vein occlusion. The beneficial effect of vitrectomy on DME can be further understood through improved oxygenation of the retina. Following vitrectomy (or a posterior

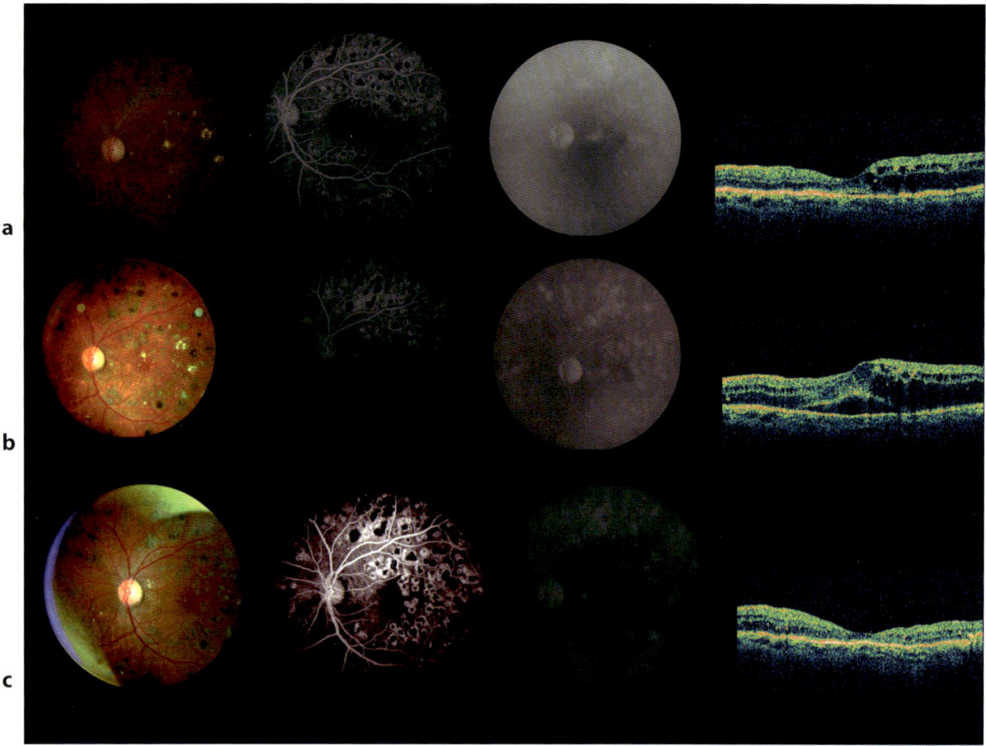

Fig. 1. **a** Color retinography, angiography and optical coherence tomography of a patient with type 2, submitted to prior laser photocoagulation and triancinolone acetonide, demonstrating clinical significant macular edema (CSME), VA 20/50. **b** Images of the 30th post-PPV plus avastin 1.25 mg showing worsening of the CSME. **c** Images of 1-year follow-up, after another two avastin injections, demonstrating regression of the edema, VA 20/25.

vitreous detachment) the ischemic retina receives oxygen from the vitreous cavity and the oxygenation of the tissue is increased. The increased oxygen level leads to reduced VEGF production and decreased vascular permeability as well as arteriolar vasoconstriction and reduced blood flow, with decreased hydrostatic pressure in the capillaries and venules and reduced edema formation. Vitrectomy alleviates hypoxia in ischemic areas of the retina which may reduce VEGF production in these areas and decreases new vessel formation [18].

Anti-VEGF Agents

A key molecule being targeted in current clinical trials is VEGF. Anti-VEGF treatments may represent as an alternative adjunctive treatment for DR and three presently available anti-VEGF agents are pegaptanib, bevacizumab and ranibizumab [19, 20].

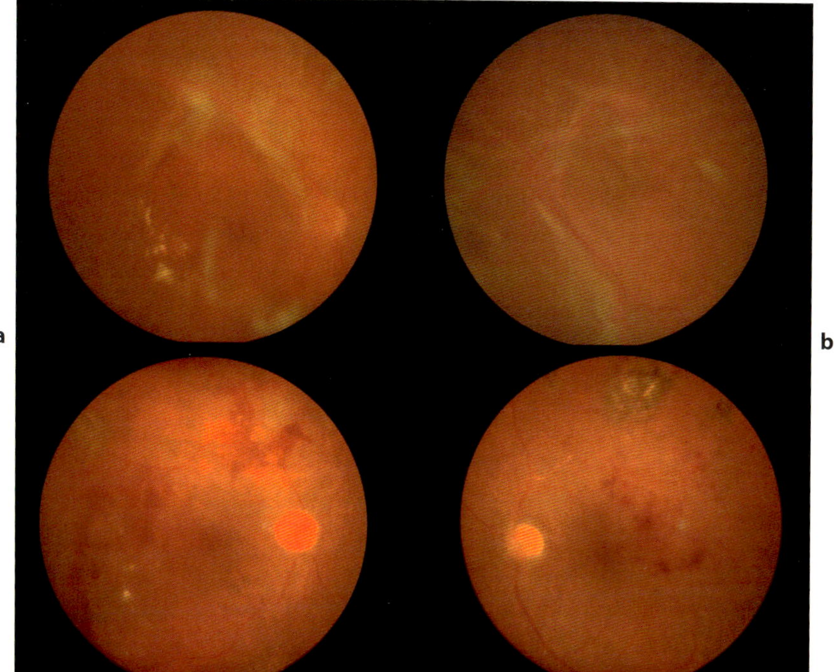

Fig. 2. Color retinography of a 32-year-old type 1 DM patient. **a, b** Right and left eye demonstrating a TRD affecting the macula. **c, d** 30th postoperative day view of right and left eye submitted to PPV plus membrane dissection and avastin 1.25 mg.

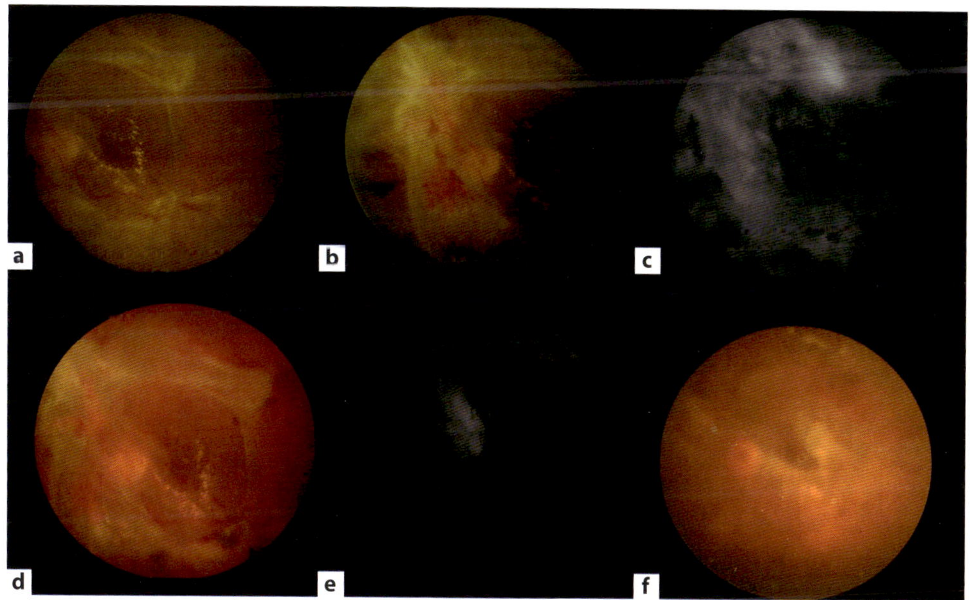

Fig. 3a–c. Color retinography and angiography of a only eye patient with very severe PDR. **d, e** Color retinography and late phase angiography 15 days after avastin injection, demonstrating involution of the fibrovascular complex. **f** 15th day post-PPV plus membrane dissection color retinography.

Pegaptanib

An inhibition of VEGF165 can be accomplished with an aptamer, a new therapeutic class of highly selective nonbiologic agents. Pegaptanib sodium is the first FDA-approved ophthalmologic anti-VEGF agent for the treatment of choroidal neovascularization (CNV) from age-related macular degeneration (AMD). Animal model studies have shown that intravitreal injection of pegaptanib can decrease the breakdown of the BRB characteristic of diabetes and can even reverse this damage to some degree. A recent phase II study reported that intravitreal pegaptanib has promising results for DME when given every 6 weeks for 30 weeks. The results showed better visual acuity at week 36 with 0.3 mg injection (20/50) as compared with sham injection (20/63) ($p = 0.04$); a mean central retinal thickness decrease of 68 μm with 0.3 mg, versus an increase of 4 μm with sham ($p = 0.02$); and reduction of subjects requiring photocoagulation in each pegaptanib arm (0.3, 1, 3 mg) [21]. Prospective clinical investigation should disclose the role of pegaptanib for the management of DME. Retrospective analysis of these data demonstrated some efficacy on retinal neovascularization as well. Phase 3 trials of pegaptanib for DME are currently being conducted.

Bevacizumab

Bevacizumab is a full-length recombinant humanized antibody active against all isoforms of VEGF-A. The US FDA approved bevacizumab in February 2004 for treatment for first-line metastatic colorectal cancer [22]. Case reports and small observational series show the efficacy of off-label intravitreal bevacizumab to treat exudative AMD [23], macular edema from nonischemic central retinal vein occlusion [24], iris neovascularization [25] and pseudophakic cystoid macular edema [26]. Haritoglou et al. [27] disclosed a prospective consecutive noncomparative case series of 51 patients treated with bevacizumab for diffuse refractory DME followed for up to 12 weeks. Their results showed a significant reduction in mean retinal thickness, from 425 μm at 2 weeks to 377 μm at 12 weeks ($p = 0.001$), although no significant changes in ETDRS letters were noticed throughout follow-up. In a second retrospective multi-center study named PACORES (Pan American Collaborative Retina Study Group), the clinical records of 88 patients with DME treated for 6 months with at least one intravitreal injection of 1.25 or 2.5 mg of bevacizumab were presented. The study found a statistically significant mean improvement in Logmar vision ($p < 0.0001$), whereas mean central macular thickness by OCT decreased over 100 μm at the end of follow-up [28]. In contrast to those results, Yanyali et al. [29] also recently showed that intravitreal bevacizumab may not promote improvement of DME or visual acuity in the short-term in 11 vitrectomized eyes. The authors attribute such findings to rapid clearance of intravitreal bevacizumab from the vitreous cavity and thus insufficient sustained therapeutic levels or individual systemic factors that may affect DME. DRCR.net has completed enrollment on a phase

2, prospective, randomized, multicenter clinical trial to determine the safety and possible benefits of this agent. Plans for a phase 3 trial of two doses of an intravitreal anti-VEGF agent versus modified ETDRS grid laser photocoagulation for DME are under discussion.

Ranibizumab

Ranibizumab is a fragment of a recombinant humanized IgG1 monoclonal antibody that inhibits all A isoforms of the human VEGF-A. Intravitreal ranibizumab is FDA approved for the treatment of exudative AMD [30]. One pilot study including 10 patients with DME, 5 receiving ranibizumab 0.3 mg and 5 receiving ranibizumab 0.5 mg, disclosed good systemic and ocular drug tolerability. At month 3, 4 of 10 patients gained >15 letters, and 8 of 10 gained >1 letter. At month 3, the mean decrease in retinal thickness of the center point of the central subfield was 45.3 ± 196.3 μm for the low-dose group and 197.8 ± 85.9 μm for the high-dose group [31]. Randomized controlled double masked trials are underway to test whether intraocular injections of ranibizumab provide long-term benefit to patients with DME. DRCR.net is planning two phase 3, prospective, randomized, multicenter trials comparing patients. In the first trial, patients with DME and no PDR will be randomized to: (1) modified ETDRS grid laser photocoagulation; (2) photocoagulation before ranibizumab; (3) photocoagulation plus IVTA, or (4) ranibizumab before photocoagulation. In the second trial, patients with DME and PDR will be randomized to: (1) modified ETDRS grid laser photocoagulation plus scatter photocoagulation; (2) modified ETDRS grid laser photocoagulation plus scatter photocoagulation plus ranibizumab, or (3) modified ETDRS grid photocoagulation plus scatter photocoagulation plus intravitreal triamcinolone injection (IVTA).

Corticosteroids and Diabetic Vitreoretinopathy

Intravitreal injection of triamcinolone acetonide (TA) may decrease DME resulting from various causes, including diabetic retinopathy 1 week to 6 months after TA injection through inhibition of VEGF cytokines. Recent prospective, randomized clinical trials have demonstrated generally favorable outcomes, the Diabetic Retinopathy Clinical Research network (DRCR.net) has completed enrollment on a three-year, randomized, prospective, multicenter clinical trial comparing two doses (1 mg and 4 mg) of preservative-free IVTA with modified early treatment diabetic retinopathy study (ETDRS) photocoagulation for DME. Long-term efficiency and safety reports of repeated intravitreal injections remain uncertain, but encouraging reports promote the development of long-term sustained release devices for corticosteroid therapy. The most important complication of IVTA is increased intraocular pressure resulting in secondary open-angle glaucoma, which sometimes may be severe and intractable.

Elevation of IOP up to 24 mm Hg may occur in about 40% of patients, usually within about 3 months. The second most important complication of IVTA is cataract formation, which may become visually significant in about half of eyes within 1 year. Other reported complications of IVTA include endophthalmitis, pseudoendophthalmitis, retinal detachment, lens trauma, and vitreous hemorrhage.

Dexamethasone is theoretically a more potent corticosteroid than TA, and intravitreal injections of dexamethasone have been shown to produce high intravitreal drug levels without toxic effects. Unfortunately, the short intraocular half-life of dexamethasone after intravitreal injection (approximately 3 h) makes this approach less useful for clinical therapy. A phase 6-month study randomized 315 patients with persistent DME with visual acuity of 20/40 to 20/200 in the study eye to observation or a single treatment with a dexamethasone implant called DDS, 350 or 700 μg. At day 90 (primary end point), an improvement in VA of 10 letters or more was achieved by a greater proportion of patients treated with dexamethasone-DDS. Results were similar in patients with diabetic retinopathy, vein occlusion, or uveitis or Irvine-Gass syndrome. During 3 months of observation, 11% of treated patients and 2% of observed patients had intraocular pressure increases of 10 mm Hg or higher. The study reported that Dexamethasone-DDS, 700 μg, may have potential as a treatment for persistent macular edema.

PPV for DR: Surgical Techniques

The diabetic retinopathy vitrectomy study (DRVS) demonstrated the efficacy of vitrectomy for severe proliferative diabetic retinopathy with non-clearing hemorrhage and progression despite laser treatment [20, 32]. At the time of the DRVS up to 20% of eyes progressed to no light perception. Today many advances have improved our success rates, including the advent of high-speed vitreous cutters, endolaser, wide-field viewing systems and pharmacological management of fibrin, retinal neovascularization and retinal edema.

The main objectives of vitrectomy are to remove media opacities, for control of active progressive PDR completely relieve all tractional adhesions, and manage recurrent complications from previous vitrectomy [33]. There are now reports in the literature suggesting that vitrectomy surgery may be helpful in eyes with refractory macular edema in eyes not responsive to laser photocoagulation.

Techniques for Membrane Dissection: Delamination, Segmentation and en-bloc Removal

The three techniques to remove the pre-retinal thickened and adherent vitreous are (1) delamination, (2) segmentation, and (3) en block dissection. (1) Delamination starts with the removal of the partially detached posterior vitreous surface between the vitreous base and the edge of the fibrovascular adhesions. Using bimanual techniques, from anterior to posterior, the edge of the fibrovascular membrane is reflected and then it can be grasped with a forceps and avulsed. (2) Segmentation is a technique used to release retinal traction caused by preretinal fibrovascular proliferation. Anteroposterior traction then is released by circumferentially cutting the posterior vitreous surface around the area of epiretinal proliferation followed by cutting of the posterior vitreous surface between epicenters of fibrovascular adhesion, leaving islands of fibrovascular tissue. (3) During an en bloc technique, the anteroposterior vitreous traction may be used to elevate the edge of the fibrovascular membrane. Initially, the vitrectomy cutter is used to create a tunnel through the formed vitreous from the sclerotomy site to an area of vitreoretinal separation. The remaining posterior vitreous surface is left attached and assists in visualizing epicenters of adhesion between fibrovascular tissue and the retina. Horizontal scissors are used to amputate epicenters of fibrovascular vitreoretinal adhesion. A better visualization of remaining cortex vitreous or adjacent epiretinal membrane may be visualized by intraoperative staining with a variety of vital dyes (Chromovitrectomy). While fluorescein stains the vitreous collagen, trypan blue may assist to visualize cellular structures of epiretinal membrane. The ILM may be stained with indocyanine green or brilliant blue [34]. After the posterior vitreous surface and the entire fibrovascular membrane are freed from the retina, they may be removed using the vitrectomy probe.

High-Speed Cutters

Several high-speed vitreous cutters are now available for diabetic vitrectomy and provide the advantage of more controlled vitreous removal and working on the retinal surface to remove complex diabetic membranes using the technique of indentation and delamination for removal of diabetic membranes and the concept of 'one-step' diabetic vitrectomy. The high-speed cutters can be used to delaminate membranes from the retinal surface and, in the majority of cases, eliminate the need for forceps, scissors, endodiathermy, healon dissection of membranes and other maneuvers traditionally used to manage these complex patients. Once one becomes experienced with this approach, cautery and most other steps may be usually eliminated.

Vitreous Base Cleaning

Scleral depression is performed while viewing through the wide-field lens system to check the retinal periphery for tears. The vitreous cutter should perform a thorough cleaning of the vitreous base to assure that: (1) postoperative retina oxygenation will be efficacious, (2) diffusion of the cytokines from ciliary body to the retina and from retina to aqueous humor will be allowed, and (3) postoperative hemorrhage will be prevented.

Small-Gauge Vitrectomy

Vitreoretinal surgeons have been developing new and innovative ways to operate with less invasive techniques. Small gauge vitrectomy instrumentation is now available allowing no-stitch surgery. 23- and 25-gauge vitrectomy instruments are now available from Bausch and Lomb, Alcon and others. These systems allow no-stitch transconjuntival insertion of vitreoretinal instrumentation, minimizing trauma to the eye and reducing the number of surgical steps. Each 25-gauge system utilizes a trocar-cannula system placed directly through the displaced conjunctiva as ports for instruments and infusion. Advantages of 25-gauge vitrectomy include a quieter eye with minimal patient discomfort and potentially less postoperative inflammation with more rapid visual recovery.

Pharmacologicaly-Assisted PPV: Anti-VEGF Agents and Corticosteroids

Diabetic Macular Edema

For DME pharmacologic agents including anti-VEGF or corticosteroids may be used intra- or postoperatively. Before surgery there is currently no rationale for the use of pharmacologic agents in DME. During surgery, the white corticosteroid triamcinolone may be injected for better visualization of the semi-transparent vitreous thereby enabling a complete vitrectomy, and this procedure has been recently called chromovitrectomy. Most patients with DR show a multilayered preretinal vitreoschisis. Therefore, when the vitreous is not stained, one or more layers of vitreous may remain, which may hamper postoperative vision improvement in patients with DME. In addition to chromovitrectomy, at the end of vitrectomy either anti-VEGF or TA may be injected into the vitreous cavity as powerful adjuvants for vitrectomy for DME.

Proliferative Diabetic Retinopathy

There is recent evidence for a benefit of preoperative anti-VEGF agents as an adjuvant to vitrectomy for the management of severe PDR [35]. Bevacizumab can potentially inhibit early neovascularization, which might be one of the causes of early recurrent hemorrhage. Then, the chances of postoperative complications such as rebleeding or fibrinoid syndrome may be decreased. It has also been hypothesized that suppression of intraocular VEGF theoretically could reduce the risk of intraoperative hemorrhage during membrane dissection facilitating the surgery. With a better visibility the surgeon may be less likely to create an iatrogenic retinal break. However, others recently proposed that even though bevacizumab may promote new vessel regression, it is unlikely to prevent early postoperative bleeding from vessels that have been injured mechanically during surgery. They suggest that the use of anti-VEGF at the end of surgery might delay the repairing process of injured blood vessels and potentially induce more recurrent hemorrhage during the early postoperative period.

New vessel formation and evolution may be affected and controlled by various growth factors and cytokines. Some factors may stimulate active vessel proliferation, while others induce fibrosis. When anti-VEGF is used, the balance among different growth factors may tip toward fibrosis inducing factors, causing contraction of the fibrovascular tissue. The manifested result may be more vitreoretinal traction or frank traction RD or even combined detachment. Arevalo et al. [36] concluded that tractional retinal detachment may occur or progress very shortly following administration of intravitreal bevacizumab in patients with severe PDR. The development or progression of tractional retinal detachment (TRD) in PDR following intravitreal bevacizumab could have happened by rapid neovascular involution with accelerated fibrosis and posterior hyaloidal contraction as a response to decreased levels of VEGF. Based on that case series and their hypothesis, the PPV should be conducted no more than seven days after the injection of anti-VEGF agents to avoid the fearful complication of TRD.

Conclusions

In DR, the gathering of preretinal serum-derived chemoattractants promotes stimulation of cellular migration on the attached posterior hyaloid. VEGF inhibitors in combination with surgical removal of the vitreous may be a reasonable approach to remove the pocket of pre-retinal growth factors and enhance the diffusion of macromolecules, including VEGF, insulin-like growth factor 1 or histamines from the retina into the vitreous. In addition, the release of tractional forces induced by the vitreomacular traction promotes additional vision improvement and reduction of macular thickness in DME. Sutureless 23–25 gauge instruments allowed the surgeon to be less aggressive to the tissues, reducing the surgical trauma and consequently less inflammation. Preoperative drug-induced fibrovascular complex regression brought the possibility to a less bleeding procedure, facilitating the dissection and reducing post-operative hemorrhage. In conclusion, new surgical techniques and pharmacological agents improved the therapeutic arsenal against this harmful disease.

References

1 Fong DS, Aiello LP, Gardner TW, et al: Retinopathy in diabetes. Diabetes Care 2004;27(suppl 1):84S-87S.
2 Preferred Practice Pattern: Diabetic Retinopathy. San Francisco, American Academy of Ophthalmology, 2003.
3 Aaberg TM, Abrams GW: Changing indications and techniques for vitrectomy in management of complications of diabetic retinopathy. Ophthalmology 1987;94:775–779.
4 Aaberg TM: Results of 100 consecutive vitrectomy procedures; in McPherson A (ed): Controversial Aspects of Vitreoretinal Surgery. St. Louis, Mosby, 1977, pp 245–249.
5 Yam JCS, Kwok AKH: Update on the treatment of diabetic retinopathy. Hong Kong Med J 2007;13:46–60.
6 Ryan SJ: Retina. St Louis, Elsevier Mosby, 2006, vol 3, pp 1–1923.
7 Meyer CH: Current treatment approaches in diabetic macular edema. Ophthalmologica 2007;221:118–131.
8 Lewis H, Abrams GW, Blumenkranz MS, Campo RV: Vitrectomy for diabetic macular traction and edema associated with posterior hyaloidal traction. Ophthalmology 1992;99:753–759.
9 Sebag J, Buckingham B, Charles MA, Reiser K: Biochemical abnormalities of vitreous in humans with proliferative diabetic retinopathy. Arch Ophthalmol 1992;110:1472–476.
10 Pendergast SD: Vitrectomy for diabetic macular edema associated with a taut premacular posterior hyaloid. Curr Opin Ophthalmol 1998;9:71–75.
11 Tachi N, Ogino N: Vitrectomy for diffuse macular edema in cases of diabetic retinopathy. Am J Ophthalmol 1996;122:258–260.
12 Lewis H: The role of vitrectomy in the treatment of diabetic macular edema. Am J Ophthalmol 2001;131:123–125.
13 Gandorfer A, Messmer EM, Ulbig MW, Kampik A: Resolution of diabetic macular edema after surgical removal of the posterior hyaloid and the inner limiting membrane. Retina 2000;20:126–133.
14 Gandorfer A, Kampik A: Surgical removal of the internal limiting membrane in diabetic cystoid macular edema. Retina 2006;26:485–486.
15 Aiello LP, Avery RL, Arrigg PG, et al: Vascular endothelial growth factor in ocular fluid of patients with diabetic retinopathy and other retinal disorders. N Engl J Med 1994;331:1480–1487.
16 Tolentino MJ, Mcleod DS, Taomoto M, et al: Pathologic features of vascular endothelial growth factor-induced retinopathy in the nonhuman primate. Am J Ophthalmol 2002;133:373–385.
17 Qaum T, Xu Q, Joussen AM, et al: VEGF-initiated bloodretinal barrier breakdown in early diabetes. Invest Ophthalmol Vis Sci 2001;42:2408–2413.
18 Stefánsson E: Ocular oxygenation and the treatment of diabetic retinopathy. Surv Ophthalmol 2006;51:364–380.
19 Emerson MV, Lauer AK: Emerging therapies for the treatment of neovascular age-related macular degeneration and diabetic macular edema. BioDrugs 2007;21:245–257.
20 Furlani BA, Meyer CH, Rodrigues EB, Maia M, Farah ME, Penha FM, Holz FG: Emerging pharmacotherapies for diabetic macular edema. Expert Opin Emerg Drugs 2007;12:591–603.
21 Cunninghan ET Jr, Adamis AP, Altawell M, et al: Macugen diabetic retinopathy study group: a phase II randomized double-masked trial of pegaptanib, an antivascular endothelial growth factor aptamer, for diabetic macular edema. Ophthalmology 2005;112:1747–1757.
22 Hurwitz H, Fehrenbacher L, Novotny W, et al: Bevacizumab plus irinotecan, fluorouracil, and leucovorin for metastatic colorectal cancer. N Engl J Med 2004;350:2335–2342.
23 Rosenfeld PJ, Moshfeghi AA, Puliafito CA: Optical coherence tomography findings after an intravitreal injection of bevacizumab (Avastin) for neovascular age-related macular degeneration. Ophthal Surg Lasers Imaging 2005;36:331–335.
24 Rosenfeld PJ, Fung AE, Puliafito CA: Optical coherence tomography findings after an intravitreal injection of bevacizumab (Avastin) for macular edema from central retinal vein occlusion. Ophthal Surg Lasers Imaging 2005;36:336–339.
25 Avery RL: Regression of retinal and iris neovascularization after intravitreal bevacizumab (Avastin) treatment. Retina 2006;26:352–354.
26 Mason JO, Albert MA, Vail R: Intravitreal bevacizumab (Avastin) for refractory pseudophakic cystoid macular edema. Retina 2006;26:356–357.
27 Haritoglou C, Kook D, Neubauer A: Intravitreal bevacizumab (Avastin) therapy for persistent diffuse diabetic macular edema. Retina 2006;26:999–1005.
28 Arevalo JF, Fromow-Guerra J, Quiroz-Mercado H, Pan-American Collaborative Retina Study Group: Primary intravitreal bevacizumab (Avastin) for diabetic macular edema: results from the Pan-American collaborative retina study group at 6-month follow-up. Ophthalmology 2007;114:743–750.
29 Yanyali A, Aytug B, Horozoglu F, Nohutcu AF: Bevacizumab (Avastin) for diabetic macular edema in previously vitrectomized eyes. Am J Ophthalmol 2007;144:124–126.

30 Heier JS, Atoszik AN, Pavan PR, et al: Ranibizumab for treatment of neovascular age-related macular degeneration: a phase I/II multicenter, controlled, multidose study. Ophthalmology 2006;113:642.E1–E4.
31 Chun DW, Heier JS, Topping TM, et al: A pilot study of multiple intravitreal injections of ranibizumab in patients with center-involving clinically significant diabetic macular edema. Ophthalmology 2006;113: 1706–1712.
32 Kroll P, Rodrigues EB, Hoerle S: Pathogenesis and classification of proliferative diabetic vitreoretinopathy. Ophthalmologica 2007;221:78–94.
33 Smiddy WE, Flynn HW Jr: Vitrectomy in the management of diabetic retinopathy. Surv Ophthalmol 1999;43:491–507.
34 Rodrigues EB, Maia M, Meyer CH, Penha FM, Dib E, Farah ME: Vital dyes for chromovitrectomy. Curr Opin Ophthalmol 2007;18:179–187.
35 Chen E, Park CH: Use of intravitreal bevacizumab as a preoperative adjunct for tractional retinal detachment repair in severe proliferative diabetic retinopathy. Retina 2006;26:699–700.
36 Arevalo F, Maia M, Flynn Jr HW, Saravia M, Avery RL, Wu L, Farah ME, Pieramici DJ, Berrocal MH, Sanchez JG: Tractional retinal detachment following intravitreal bevacizumab (Avastin) in patients with severe proliferative diabetic retinopathy Br J Ophthalmol 2008;92:213–216.

Carsten H. Meyer, MD
Department of Ophthalmology, University of Bonn
Ernst-Abbe-Strasse 2
DE–53127 Bonn (Germany)
Tel. +49 228 2871 11515, Fax +49 228 2871 5603, E-Mail meyer_eye@yahoo.com

Gandorfer A (ed): Pharmacology and Vitreoretinal Surgery.
Dev Ophthalmol. Basel, Karger, 2009, vol 44, pp 82–88

Tissue Plasminogen Activator-Assisted Vitrectomy: Surgical Drainage of Submacular Hemorrhage

Motohiro Kamei · Yasuo Tano

Department of Ophthalmology, Osaka University Graduate School of Medicine, Suita, Japan

Abstract

A thick subretinal hemorrhage involving the macula is currently treated with pneumatic displacement or surgical removal. We evaluated the long-term effects of tPA-assisted surgical drainage of a submacular hemorrhage by reviewing the medical records of 12 eyes with submacular hemorrhage followed for a mean follow-up time of 6.9 years. The final best corrected visual acuity (BCVA) improved in 10 eyes (83%), was unchanged in 2 (17%), and was reduced in 0 (0%). The mean preoperative BCVA was 6/200, and the best postoperative BCVA was 20/47, and the final BCVA was 20/143. The late postoperative complications, including recurrent submacular hemorrhage, enlargement of the CNV, and retinal degeneration, reduced the final BCVA. These results indicate that tPA-assisted surgical removal of a subretinal hemorrhage can lead to improved long-term vision. Controlling the choroidal neovascularization may be important for retaining the best postoperative vision.

The visual prognosis of eyes with a submacular hemorrhage associated with age-related macular degeneration (AMD) is generally poor especially when the clot is thick and large [1]. The permanent visual loss after a subretinal hemorrhage is due to photoreceptor damage caused by metabolic interference, iron toxicity, and the shearing of the photoreceptors by a contraction of the fibrin. Experimental studies have shown that the fibrin-mediated damage to the outer retina occurs within 24 h of the onset [2]. Therefore, a thick subretinal hemorrhage involving the macula is currently treated with pneumatic displacement or surgical removal as soon as possible. Mechanical removal or displacement, however, may cause a shearing of the photoreceptors because fibrin interdigitates between the hemorrhagic clot and the photoreceptors.

To overcome this problem, a simultaneous use of tissue plasminogen activator (tPA) is therefore recommended. tPA is a recombinant serine protease which lyses serine residues in plasminogen and activates it to plasmin resulting in fibrinolysis. Although several studies have reported on the short-term efficacy of tPA-assisted surgical drainage of submacular hemorrhage [3–8], there is no report on the long-term results.

We report on the visual outcome of 12 patients who underwent tPA-assisted surgical drainage of submacular hemorrhage associated with AMD and were followed for more than 5 years after the surgery.

Patients and Methods

Patients

We reviewed the medical records of 37 eyes with submacular hemorrhage that had undergone tPA-assisted surgical drainage at Osaka National Hospital from July 1992 to October 1996. From the 37 eyes, we selected 12 eyes which were followed for more than 5 year postoperatively. The preoperative ophthalmic examination included measurements of the visual acuity and intraocular pressure, slit-lamp and indirect ophthalmoscopic examination, visual field analyses, and fluorescein angiography.

Surgical Procedures

A standard three-port pars plana vitrectomy was performed, and the posterior hyaloid membrane was carefully separated from the retina. A bent tip 27- or 30-gauge needle (blunt or sharp) was inserted into the subretinal hemorrhage at its superotemporal edge without prior retinotomy or endodiathermy. Then 25 μg/0.1 ml of tPA was slowly injected until it covered the submacular hemorrhage. The total volume of injected tPA ranged from 0.05 and 0.2 ml, which corresponds 12.5–50 μg, depending on the volume of the subretinal blood. After waiting 40–60 min for fibrinolysis to occur, the injection site was enlarged to approximately 1/2 disc diameter by retinotomy with vitreoretinal scissors. Perfluorocarbon liquid was injected onto the inferonasal retina until it reached the retinotomy. The liquefied subretinal hemorrhage was then evacuated into the vitreous cavity through the retinotomy and aspirated with a silicone-tipped back flush needle. A rocking motion to the eye was often used to squeeze the subretinal hemorrhage. Balanced saline solution was injected slowly into the subretinal space through the retinotomy in order to dilute and flush out the fibrin degradation products (FDPs) which were generated from fibrinolysis by plasmin activated by tPA. The FDPs were removed because they have strong chemotactic and inflammatory properties. After a second administration of liquid perfluorocarbon and manipulations to squeeze out the FDPs, a fluid-air exchange was performed. The operation was concluded with infusion of 20% SF_6 gas.

Examinations

Postoperatively, the patients underwent the same examination as preoperatively under the same conditions. The best-corrected visual acuity (BCVA) was examined by technicians who were masked to the surgical procedures. Changes of the BCVA were assessed according to the postoperative times. Postoperative complications were also recorded and evaluated.

Results

There were 8 men and 4 women whose average age at the time of surgery was 62.2 years. The follow-up period ranged from 5 to 10 years (mean: 6.9 years). The final visual acuity improved by 2 or more lines from the preoperative vision in 10 eyes (83%), was unchanged in 2 (17%), and was reduced in none. Three eyes (25%) had a final BCVA of ≥20/30, 4 eyes (33%) had 20/200 to 20/40, and 5 eyes (42%) had <20/200 visual acuity. The mean preoperative BCVA was 6/200, the postoperative best BCVA was 20/47, and the final BCVA was 20/143.

The early postoperative complications included retinal detachment in one eye (8%) and progression of cataract in 9 eyes (75%). The vision in the eyes with these complications recovered following surgical management. The late postoperative complications included recurrent submacular hemorrhage in 5 eyes (42%), CNV enlargement in 4 eyes (33%), epiretinal membrane formation in 3 eyes (25%), and progression of chorioretinal atrophy in 2 eyes (17%). These late postoperative complications reduced the postoperative BCVA. The postoperative BCVA was maintained in only 4 eyes (33%). Time courses of visual acuity in representative cases are shown in figures 1–4.

Discussion

Bennett et al. [1] reported that the mean final visual acuity in eyes with a submacular hemorrhage was 20/1,700. In another clinical study describing the natural history of submacular hemorrhage associated with AMD, retinal arterial macroaneurysms, and ocular histoplasmosis syndrome, the percentage of eyes that had a significant visual recovery was <50% [9]. Compared to the natural course of a submacular hemorrhage, our results in our 12 patients were better, i.e. they had a mean final visual acuity of 20/143 at almost 7 years after the surgery and a visual recovery rate of 83%.

A number of studies on the short-term results of surgically removing a submacular hemorrhage reported a visual improvement in >70% of the eyes [3–8, 11]. As best we know, there is no report on the long-term results of the visual outcome after a surgical removal of a submacular hemorrhage. Our results demonstrated that tPA-assisted surgical removal of submacular hemorrhage had a long-term effect with an average of approximately 7 years. The long-term results in our patients showed that the visual acuity improved in 83% and the final visual acuity was 20/200 or better in 58%. These findings are comparable to our earlier study of the short-term results where we found that the visual acuity improved in 72 of 92 eyes (78%) and 56% of the eyes had a final visual acuity of ≥20/200.

The surgical removal of a subretinal hemorrhage by forceps was first reported by Hanscom and Diddie [10] in 1987. The mechanical removal of a clot was associated with poor visual results probably because the fibrin interdigitates between the hemorrhagic clot and the photoreceptors, and the removal of the clot resulted in significant damage to the outer retina. Experimental studies have shown that the fibrin-mediated irreversible damage to the outer retina occurs within 24 h of onset [2]. The use of pharmacological agents to release fibrin from the photoreceptor outer segments

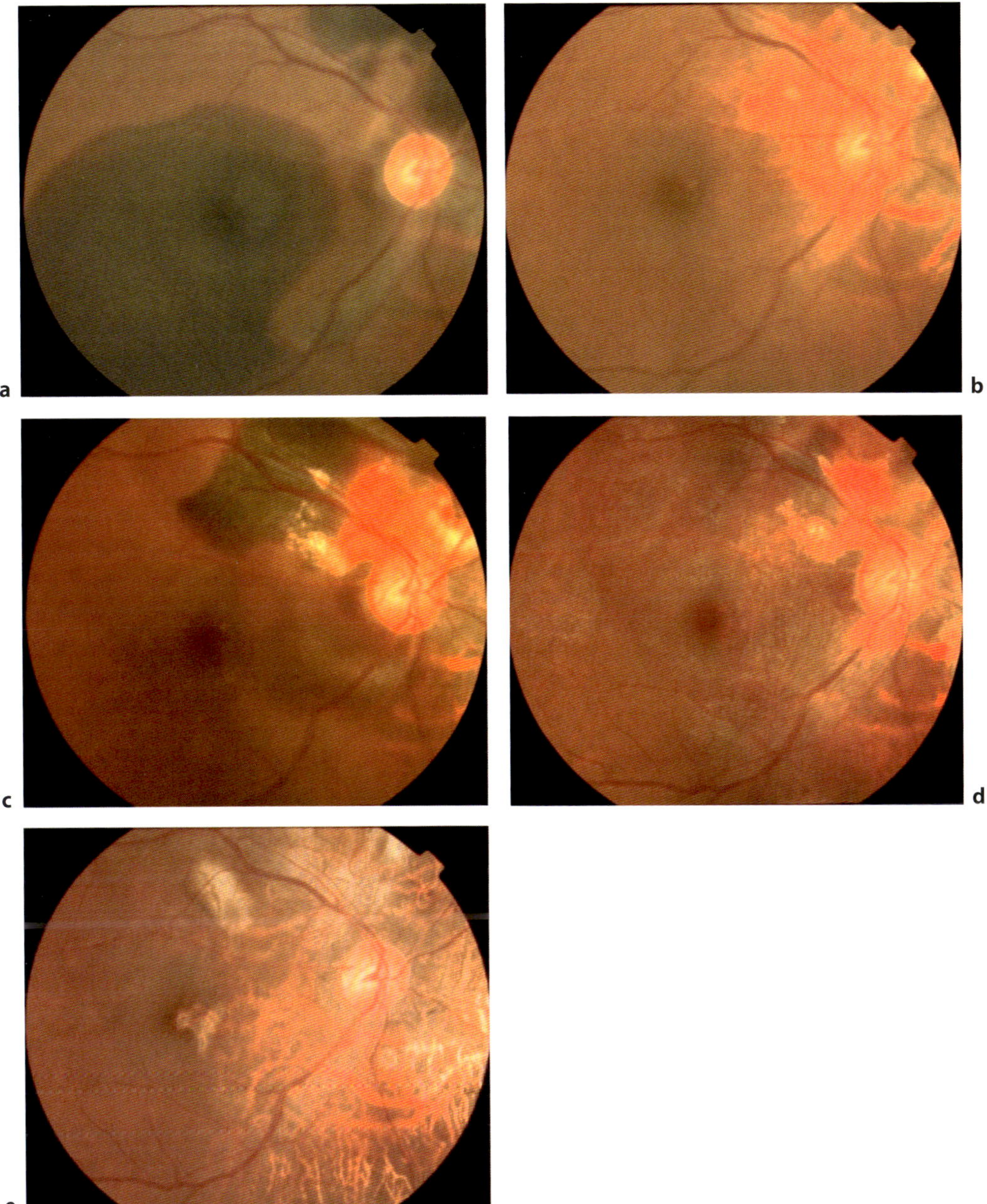

Fig. 1. Series of fundus photographs of a representative case (71-year-old man). Thick submacular hemorrhage (**a**) decreased the vision to counting fingers which improved to 20/25 with tPA-assisted SRH drainage surgery (**b**: 6 months after the first surgery). Subretinal hemorrhage recurred at 1.5 years after the first surgery and decreased his vision to 4/200 (**c**). The subretinal hemorrhage was successfully removed again at second surgery and vision recovered to 20/30 (**d**). Vision has been maintained for more than 4 years after the last surgery (**e**).

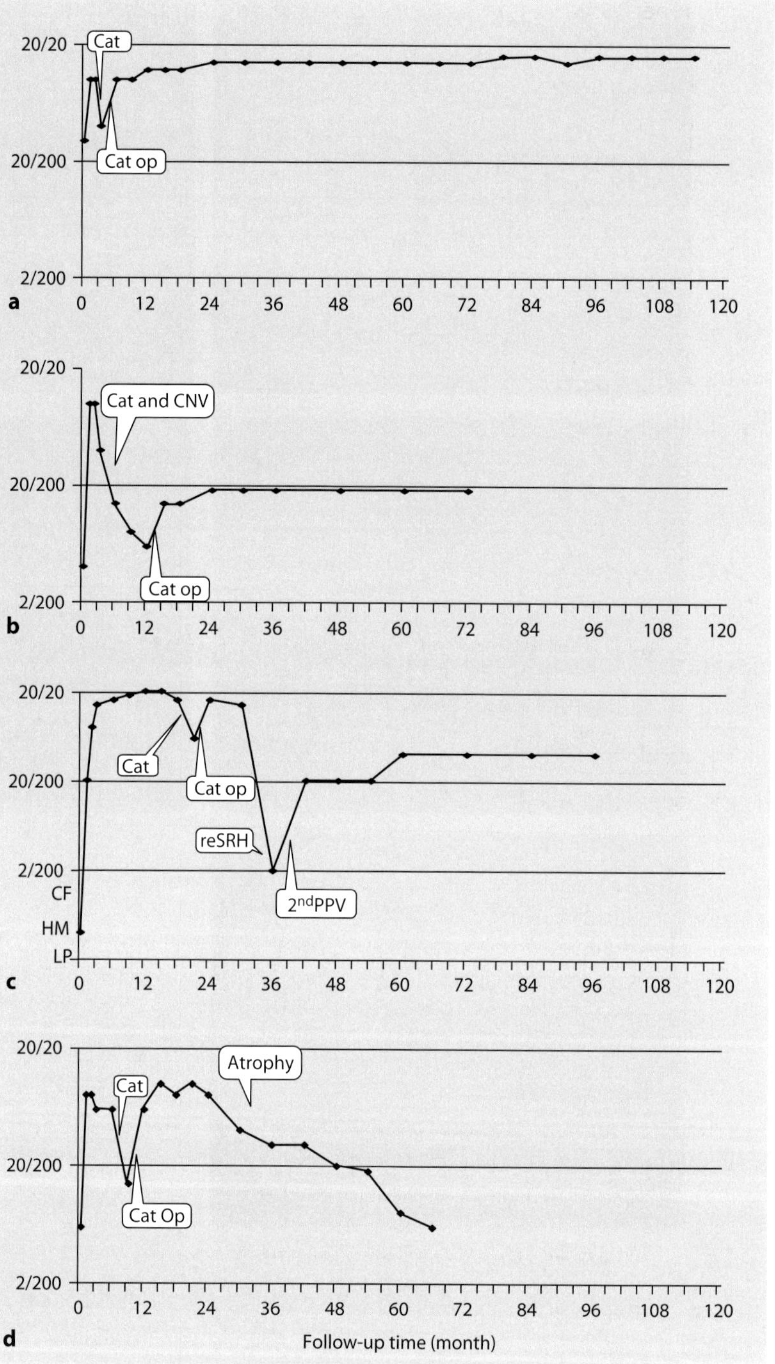

Fig. 2. Time courses of visual acuity in representative cases. **a** Case 1: a 63-year-old man. His cataract worsened in the 3 months after the surgery, resulting in a decrease of vision. His vision improved up to 20/25 after cataract surgery and was maintained for more than 9 years after the first surgical removal of a submacular hemorrhage. **b** Case 2. a 70-year-old woman. Her best visual acuity improved to 20/40 immediately after the surgery, but cataract progression and CNV enlargement depressed her vision. Cataract surgery 1 year after the vitrectomy partly improved the vision, but unfortunately remained low at 18/200. **c** Case 3: a 76-year-old man. Cataract progressed 2 years after the surgery resulting in a decrease of vision. His vision improved up to 20/25 after cataract surgery, but a recurrence of a submacular hemorrhage occurred 3 years after the surgery and the visual acuity decreased to 2/200. The second surgical removal of submacular hemorrhage was performed and the vision recovered up to 20/100, which was maintained for 5 years after the last surgery. **d** Case 4: a 74-year-old woman. A good visual acuity of 20/40 was maintained for 2 years with cataract surgery. However, chorioretinal atrophy gradually progressed and decreased the vision to 6/200.

appears reasonable and necessary. Subretinal injections of tPA have been shown to liquefy submacular hemorrhages, and the liquefied clots are evacuated through a small retinotomy, which reduced surgery-induced damage to the neural retina and retinal pigment epithelium.

The early postoperative complications of surgical removal of submacular hemorrhage in the previous reports [3–8] included disciform scar formation due to persistent choroidal neovascularization (11%), recurrence of the hemorrhage (5–18%), retinal detachment and PVR (4–14%), epiretinal membrane formation (14%), and cataract progression (12–53%). In our patients, we found retinal detachment in one eye (8%) and progression of cataract in 9 eyes (75%) as early postoperative complications, and recurrent submacular hemorrhage in 5 eyes (42%), CNV enlargement in 4 eyes (33%), epiretinal membrane formation in 3 eyes (25%), and progression of chorioretinal atrophy in 2 eyes (17%) as late postoperative complications. Because these late postoperative complications depressed the postoperative BCVA, suppressing CNV activity seems important for maintaining the better vision.

Current Indications for tPA-Assisted Surgical Removal of Submacular Hemorrhage

Although the first choice for treating subretinal hemorrhage is currently pneumatic displacement, we demonstrated in this study that tPA-assisted surgical removal can lead to improved long-term vision. We are currently using tPA-assisted surgical removal of submacular hemorrhage in the following cases: submacular hemorrhage that lasts less than 2 weeks or appears predominantly red with minimal yellowish-white color, ≥500 μm thick, and ≥2 disc area in size. Other factors, such as conditions of the opposite eye, systemic conditions, and visual acuity immediately before the submacular hemorrhage developed, should be considered for selecting tPA.

References

1 Bennett SR, Folk JC, Blodi CF, Klugman M: Factors prognostic of visual outcome in patients with subretinal hemorrhage. Am J Ophthalmol 1990;109:33–37.

2 Toth CA, Morse LS, Hjelmeland LM, Landers MB 3rd: Fibrin directs early retinal damage after experimental subretinal hemorrhage. Arch Ophthalmol 1991;109:723–729.

3 Lewis H: Intraoperative fibrinolysis of submacular hemorrhage with tissue plasminogen activator and surgical drainage. Am J Ophthalmol 1994;118:559–568.

4 Ibanez HE, Williams DF, Thomas MA, Ruby AJ, Meredith TA, Boniuk I, Grand MG: Surgical management of submacular hemorrhage: a series of 47 consecutive cases. Arch Ophthalmol 1995;113:62–69.

5 Lim JI, Drews-Botsch C, Sternberg P Jr, Capone A Jr, Aaberg TM Sr: Submacular hemorrhage removal. Ophthalmology 1995;102:1393–1399.

6 Kamei M, Tano Y, Maeno T, Ikuno Y, Mitsuda H, Yuasa T: Surgical removal of submacular hemorrhage using tissue plasminogen activator and perfluorocarbon liquid. Am J Ophthalmol 1996;121:267–275.

7 Moriarty AP, McAllister IL, Constable IJ: Initial clinical experience with tissue plasminogen activator (tPA) assisted removal of submacular haemorrhage. Eye 1995;9:582–588.
8 Kamei M, Estafanous M, Lewis H: Tissue plasminogen activator in the treatment of vitreoretinal diseases. Semin Ophthalmol 2000;15:44–50.
9 Berrocal MH, Lewis ML, Flynn HW Jr: Variations in the clinical course of submacular hemorrhage. Am J Ophthalmol 1996;122:486–493.
10 Hanscom TA, Diddie KR: Early surgical drainage of macular subretinal hemorrhage. Arch Ophthalmol 1987;105:1722–1723.
11 Shiraga F, Matsuo T, Yokoe S, Takasu I, Okanouchi T, Ohtsuki H, Grossniklaus HE: Surgical treatment of submacular hemorrhage associated with idiopathic polypoidal choroidal vasculopathy. Am J Ophthalmol 1999;128:147–154.

Motohiro Kamei, MD
Department of Ophthalmology, Osaka University Medical School
2–2 Yamadaoka
Suita, 565-0871 (Japan)
Tel. +81 6 6879 3456, Fax +81 6 6879 3458, E-Mail mkamei@ophthal.med.osaka-u.ac.jp

Gandorfer A (ed): Pharmacology and Vitreoretinal Surgery.
Dev Ophthalmol. Basel, Karger, 2009, vol 44, pp 89–97

Anti-Angiogenic Therapy in the Management of Retinopathy of Prematurity

Kimberly A. Drenser

Associated Retinal Consultants, Royal Oak, Mich., and Eye Research Institute, Oakland University, Rochester, Mich., USA

Abstract

Retinopathy of prematurity (ROP) is a vitreoretinal abnormality that significantly affects premature babies with low birth rates. Despite improved screening and management of these infants, a subset will progress to retinal detachment and permanent visual impairment. Current treatment consists of peripheral laser ablation and subsequent surgical intervention if a detachment occurs. We sought to evaluate the vitreous biochemistry of eyes that progress despite appropriate laser intervention. Additionally, a limited trial of an anti-VEGF (vascular endothelial growth factor) therapy was used in one eye of infants with persistent Plus disease and neovascularization. The anti-VEGF treatment successfully decreases abnormal angiogenesis but does not decrease the proliferative changes associated with retinal detachment. Biochemical analysis of the vitreous of stage 4 ROP eyes shows significantly elevated VEGF and transforming growth factor (TGF-β) concentrations, and normal levels of other angiogenic factors.

Background

Retinopathy of prematurity (ROP) continues to be a leading cause of blindness in children in developed countries around the world, and an increasing cause of blindness in developing countries. The current standard of treatment is ablation of the peripheral avascular retina. Screening and intervention is now directed by the criteria established by the ETROP Study [1]. The ETROP Study demonstrated superior results compared to the Cryo-ROP Study but also recommends earlier treatment with laser ablation. In the children requiring laser treatment the peripheral retina is ablated and destroyed for future use. The ablated retina is not functional and is not amenable to regeneration. In addition, morbidity from laser can include cataract and vitreous hemorrhage as well as anterior segment ischemia, which can lead to phthisis bulbi.

The mechanism driving the development of ROP is in large part dependent on vascular endothelial growth factor (VEGF) [2–5]. The normal biochemistry of the developing eye is altered due to the change in environment when a baby is premature. During the second trimester the developing fetus enters phase 1 of vascular development. During phase 1 the endogenous VEGF levels of the fetus are elevated to promote angiogenesis and vascular maturation. In contrast, the relatively hyperoxic environment that the premature baby is introduced in to results in the converse reaction, with decreased production of VEGF leading to delayed retinal maturation. Phase 2 occurs during the third trimester and during this phase the developing fetus will normally reduce VEGF production [6]. VEGF production becomes dysregulated in the premature child, presumably due to large areas of avascular retina creating tissue hypoxia, resulting in abnormally high levels of VEGF. This biochemical shift heralds the pathological changes seen in ROP and explains why the peak incidence of stage 3 ROP occurs between a postmenstrual age of 36–38 weeks [3].

Recently, a subset of patients developing a severe form of ROP has been identified [1]. This group develops an aggressive form of zone 1 disease, designated as aggressive posterior ROP (APROP). APROP develops between postmenstrual age (PMA) 32–34 weeks and is seen in very low birth weight infants generally with a GA less than 26 weeks. It has the poorest outcome with many eyes progressing to retinal detachment despite complete and timely laser ablation of the avascular retina [7].

Vitreous Biochemistry

Vitreous samples were taken at the time of surgery for stage 4 ROP eyes (± pegaptinib treatment). The fluid was analyzed by ELISA assay for total VEGF, TGF-β (isomers 1 and 2), hepatocyte growth factor (HGF) and insulin growth factor (IGF-1). We found significantly elevated levels of VEGF and TGF-β in eyes with ROP-related retinal detachment (fig. 1). The average age of surgery was postmenstrual age 42 weeks, a time point that represents the most frequent onset of tractional retinal detachment in ROP eyes that have undergone laser ablation. Vitreous VEGF concentrations were 576 pg/mg protein in vascularly active eyes and 56 pg/mg protein in nonactive ROP eyes, compared to 12 pg/mg protein in control eyes. The VEGF was most elevated in eyes that demonstrated persistent vascular activity characterized as Plus disease or neovascularization at the time of surgery [2]. It is interesting that despite appropriate and timely laser ablation elevated levels of VEGF persist in the ROP eyes. This finding indicates that the biochemical disturbance is not due to avascular retina alone.

There are a number of other angiogenic factors that have been associated with neovascularization in various eye diseases. HGF and IGF-1 have been associated with proliferative diabetic retinopathy [8–10], although there have been discrepancies regarding the intravitreal HGF levels [11]. We evaluated the vitreous samples for other factors implicated in angiogenesis (HGF and IGF-1) and found that they were not elevated in eyes with ROP compared to controls. From these studies, it appears that VEGF is the primary factor driving Plus disease and neovascularization in ROP.

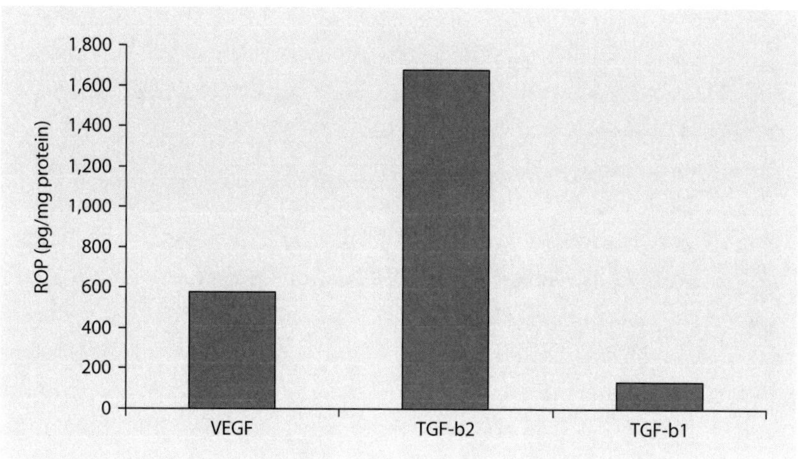

Fig. 1. Graphic depiction of intravitreal growth factor concentrations in stage 4 ROP.

Transforming growth factor has an anti-angiogenic activity and also increases preretinal proliferative changes. It has been noted that TGF-β levels increase near the due date of premature infants and may play a role in the final suppression of intraocular angiogenesis. The majority of infants with ROP requiring laser ablation do not go on to develop a retinal detachment. Interestingly, those infants that do develop a tractional retinal detachment following appropriate laser ablation do so shortly after their due date (highest incidence at PMA 42 weeks). The vitreous samples of stage 4 ROP eyes were evaluated for the concentrations of the two predominant intraocular isomers of TGF-β (1 and 2). TGF-β2 is the predominant form, accounting for nearly 90% of the total TGF-β. Although TGF-β1 only accounts for approximately 10%, it may play a greater role in proliferative events due to its control of pericyte growth [12]. TGF-β1 averaged 145 pg/mg protein in ROP eyes compared to 4 pg/mg protein in control eyes. TGF-β2 averaged 1,744 pg/mg protein in ROP eyes compared to 470 pg/mg protein in control eyes. When compared to controls these are significantly higher concentrations. However, the TGF-β levels for both isoforms are also significantly higher than those seen in rhegmatogenous retinal detachment and proliferative diabetic retinopathy (fig. 2, 3). This finding supports the suggestion that TGF-β levels increase with increasing elevations of VEGF and likely accounts for the small subset of ROP eyes that progress to tractional retinal detachment despite timely and complete laser ablation of the avascular retina.

Anti-Angiogenesis Treatment

With the advent of FDA approved drugs for anti-VEGF treatment, the possibility of treating ROP with anti-VEGF agents has become possible. Drugs that are available

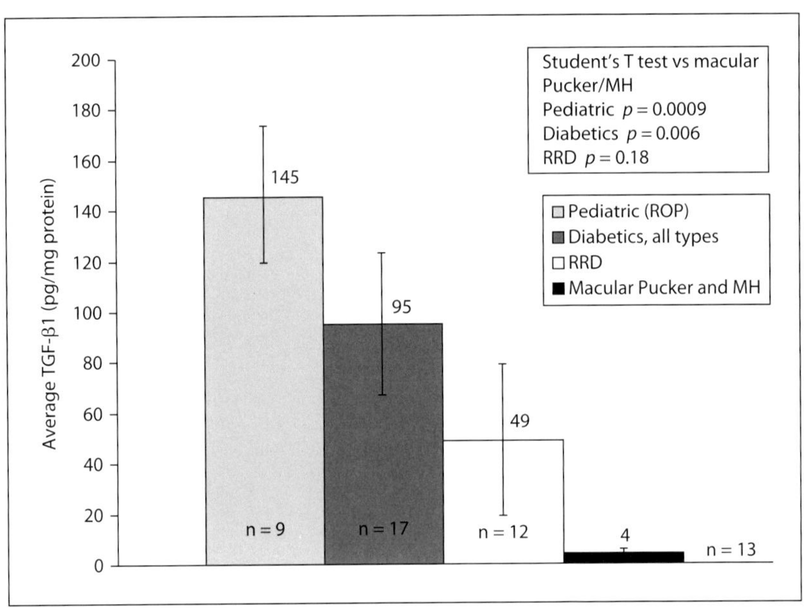

Fig. 2. Graphic depiction of TGF-β1 concentrations.

include the drug pegaptinib (Macugen) for partial blockage of VEGF-A, or complete blockage of VEGF-A with drugs such as ranibizumab (Lucentis) and bevacizumab (Avastin). Some clinicians have evaluated the usefulness of bevacizumab (Avastin) in ROP. These were retrospective case reports or series. All reports demonstrated safety and tolerance of an intravitreal anti-VEGF treatment and all investigators found a rapid resolution of Plus disease and/or neovascularization [13, 14]. One report, however, did find severe progression of fibrosis and retinal detachment with a funnel detachment ensuing [15]. Of note, this particular case received the treatment when proliferative changes were noted and an early detachment was already occurring. Many clinicians have seen similar tractional changes following anti-VEGF treatment of patients with proliferative diabetic retinopathy [16, 17].

In a prospective limited trial, our institution evaluated the effect of partial VEGF inhibition using pegaptinib. VEGF is required in the developing retina for normal angiogenesis and neurogenesis [4] and it was felt that partial VEGF blockade may be beneficial to the developing retina and its vasculature. Pegaptinib was developed to inhibit the $VEGF_{163}$ isomer only, with the theory that this particular isomer represents the pathological changes seen with increased VEGF concentrations. The goal of intravitreal anti-VEGF treatment was to block the excessive levels of VEGF trapped within the overlying vitreous and maintain a normal vanguard of intraretinal VEGF, promoting continued vascular maturation. Eyes requiring surgery had biochemical analysis of their vitreous fluid in order to evaluate other potentially important agents in the development of ROP and progression of retinal detachment.

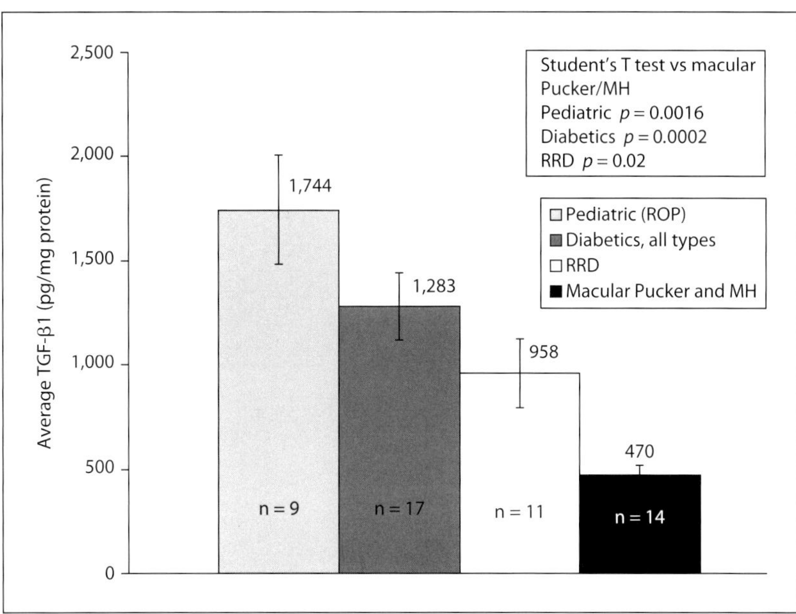

Fig. 3. Graphic depiction of TGF-β2 concentrations.

Premature babies with zone 1 ROP (fig. 4) requiring treatment per ETROP criteria underwent standard care using peripheral laser ablation. Infants with zone 1 disease who demonstrated continued vascular abnormalities (persistent Plus disease or neovascularization) despite appropriate and complete laser ablation and were less than 36 weeks of gestational age were offered rescue therapy with pegaptinib. A total of 5 infants participated and each infant had one eye treated with intravitreal anti-VEGF agent. The study eye received two injections of 1/2 dose pegaptinib (0.05 ml) 1 week apart. The fellow eye was managed by standard care with laser ablation and surgical intervention if progression to tractional retinal detachment was noted. Dilated fundus examination with fundus photography was performed weekly. Surgery was performed if a tractional retinal detachment developed in either eye.

Three infants progressed to bilateral retinal detachment despite treatment (laser ablation of both eyes; one eye received pegaptinib). Of note, all of the pegaptinib-treated eyes demonstrated prompt resolution of Plus disease and regression of the tunica vasculosa lentis 1 week after the initial intravitreal injection (fig. 5) (1/2 dose), consistent with the findings of other investigators. The eyes remained quiet for the remainder of the study. Also, of interest, the pegaptinib-treated eyes that went on to tractional retinal detachment developed the detachment 1–2 weeks after the laser – only eye developed detachment, indicating that decreasing vascular permeability delays the onset of retinal detachment. Once retinal detachment began, however,

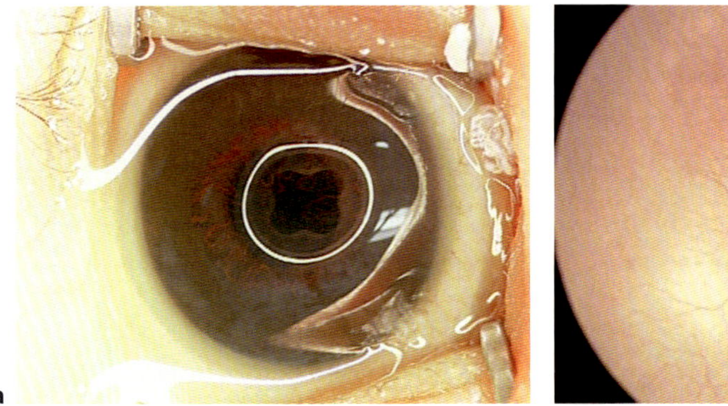

Fig. 4. **a** Persistent tunica vasculosa lentis with poor pupil dilation. **b** Zone 1, stage 3, early Plus. Both eyes had symmetrical disease.

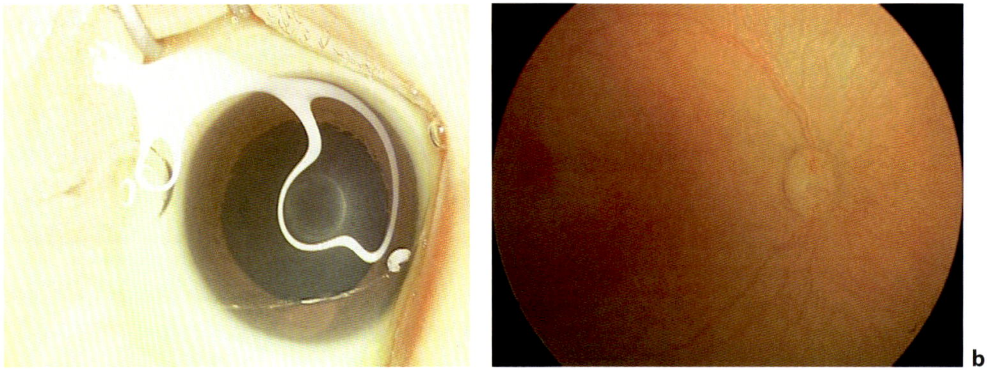

Fig. 5. Zone 1, stage 3, early Plus OU; 1 week after treatment. **a** Right eye: treated with laser ablation and pegaptinib. Note resolution of Plus disease. **b** Left eye: treated with laser only. Note persistent Plus disease.

the pegaptinib-treated eyes had an aggressive fibrosis and proliferation that led to an accelerated detachment progression (fig. 6). Despite the anti-VEGF treatment not preventing retinal detachment, early surgical intervention at the first sign of fibrosis and/or retinal detachment leads to an improved long-term result, both structurally and functionally, compared to the laser-only fellow eye (fig. 7).

The two pegaptinib-treated eyes that did not progress to tractional retinal detachment received the first intravitreal injection at the time of laser, and therefore, at the youngest age (post-menstrual age 33 weeks) in this study group. Presumably, this early anti-VEGF inhibition results in early reversal of the pathologic effects of VEGF accumulation in the vitreous and stabilizes the intraocular biochemistry. Blockage of VEGF accumulation when the fetus is entering phase 2 may improve the anti-VEGF treatment response and reverse the pathological angiogenesis.

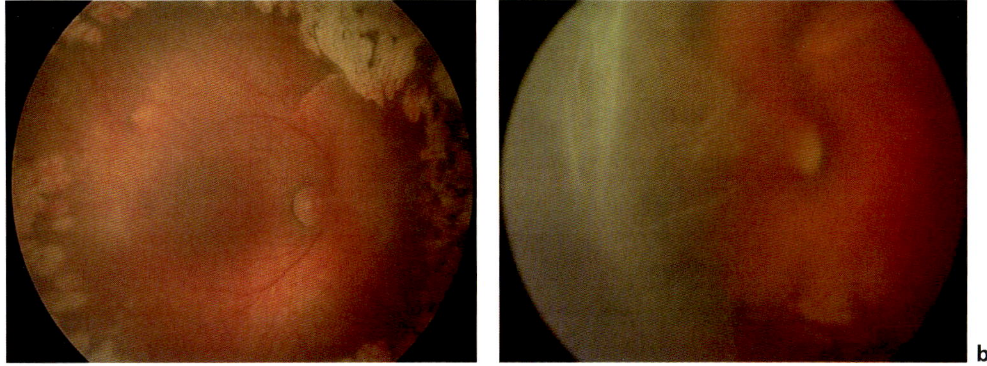

Fig. 6. Postmenstrual age 42 weeks; bilateral retinal detachment. **a** Right eye: early nasal retinal detachment with posterior pole distortion s/p pegaptinib. **b** Left eye: extensive nasal retinal detachment with folding of the temporal retina.

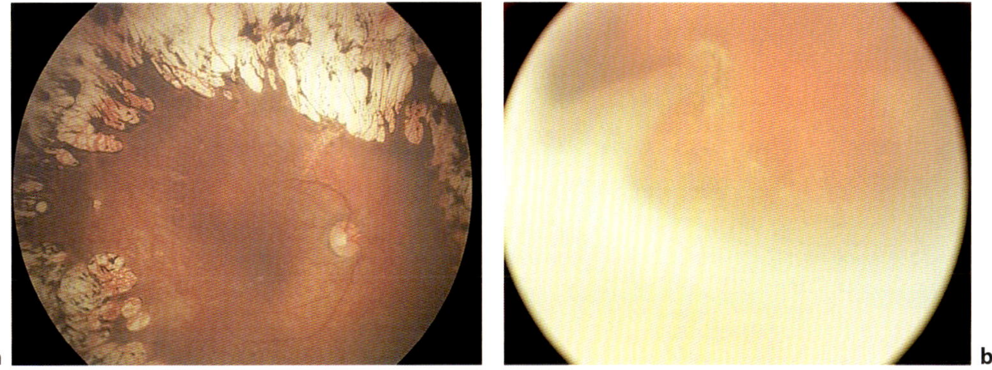

Fig. 7. 2 months s/p vitrectomy OU. **a** Right eye: excellent anatomical and visual outcome (20/40). **b** Left eye: poor anatomical outcome with light perception only.

As noted previously, it has been suggested that TGF β levels endogenously upregulate near the due date and that the level of TGF-β corresponds to the levels of VEGF present at that time. Binding of VEGF does not appear to alter the concentration or effects of TGF-β. TGF-β plays a role in proliferative changes affecting the retina and the rapid binding of VEGF may allow unopposed TGF-β to have a more dramatic proliferative response. Support of this theory is seen in the vitreous sample of infants with stage 4 ROP, where the levels of TGF-β are elevated significantly above those seen in other proliferative diseases (fig. 2, 3). Additionally, the intravitreal concentrations of TGF-β are greatly elevated compared to the VEGF concentrations in eyes with stage 4 ROP at the time of surgery (fig. 1). These findings indicate that early VEGF inhibition is key to successful anti-VEGF treatment in ROP.

Conclusion

The biochemical constituents of patients with stage 4 ROP indicate that abnormally high levels of VEGF and TGF-β are present and appear to be the predominant factors driving pathological angiogenesis and retinal detachment. HGF and IGF-1 have been implicated in angiogenesis, especially in diabetic patients [9, 10, 18]. HGF has also been implicated in neuronal differentiation [11, 19] and is lower than expected in these samples. Both HGF and IGF-1 are expressed at low levels and do not appear to play a significant role in ROP pathology. The surprisingly low levels of IGF-1 in the vitreous of ROP infants may be an indication of delayed infant maturation or further biochemical abnormalities in the premature eye.

Inhibition of VEGF effectively decreases the vascular pathology and results in stabilization of the vasculature [4]. It does not, however, prevent the progression to retinal detachment. The significantly elevated levels of TGF-β likely contribute to the tractional changes seen in theses eyes. TGF-β may be abnormally elevated as a secondary response to high intravitreal VEGF concentrations. Possibly, the blocked VEGF leads to unopposed TGF-β activity and promotes proliferation of cells and tractional retinal detachment. Additionally, the incomplete regression of primary vasculature and vitreous may provide a platform of increased cellular substrate for TGF-β to act upon, creating greater tractional forces. Earlier intervention with an anti-VEGF agent may avoid this late complication and further studies will need to be conducted to answer this. Prompt surgical intervention, however, results in a favorable outcome. The anti-VEGF treatment quiets the vascular abnormalities and creates less intraoperative bleeding, allowing for early surgical intervention if necessary. Alternatively, a TGF-β inhibitor may be helpful in preventing retinal detachment in these patients.

Continued investigation into the pathology of ROP and the use of biochemical modulation with timely surgical intervention may lead to improved visual outcomes for these infants.

References

1 Early Treatment for Retinopathy of Prematurity Cooperative Group: Revised indications for the treatment of retinopathy of prematurity: results of the early treatment for retinopathy of prematurity randomized trial. Arch Ophthalmol 2003;121:1684–1694.

2 Sonmez K, Drenser KA, Capone A Jr, Trese MT: Vitreous levels of stromal cell-derived factor-1 and vascular endothelial growth factor in patients with retinopathy of prematurity: implications for vascular activity. Ophthalmology 2008;115:1065–1070.

3 Stone J, Itin A, Alon T, et al: Development of retinal vasculature is mediated by hypoxia-induced vascular endothelial growth factor (VEGF) expression by neuroglia. J Neurosci 1995;15:4738–4747.

4 Hartnett ME, Martiniuk D, Byfield G, Geisen P, Zeng G, Bautch VL: Neutralizing VEGF decreases tortuosity and alters endothelial cell division orientation in arterioles and veins in a rat model of ROP: relevance to plus disease. Invest Ophthalmol Vis Sci 2008;49:3107–3114.

5 Pierce EA, Avery RL, Foley ED, et al: Vascular endothelial growth factor/vascular permeability factor expression in a mouse model of retinal neovascularization. Proc Natl Acad Sci USA 1995;92: 905–909.
6 Smith LE: Retinopathy of prematurity. Angiogenesis 2007;10:133–140.
7 Kychenthal A, Dorta P, Katz X: Zone 1 retinopathy of prematurity: clinical characteristics and treatment outcomes. Retina 2006;26:S11–S15.
8 Nowak M, Wielkoszyński T, Marek B, Kos-Kudła B, Swietochowska E, Siemińska L, Kajdaniuk D, Głogowska-Szelag J, Nowak K: A comparison of the levels of hepatocyte growth factor in serum in patients with type 1 diabetes mellitus with different stages of diabetic retinopathy. Endokrynol Pol 2008; 59:2–5.
9 Simó R, Carrasco E, García-Ramírez M, Hernández C: Angiogenic and antiangiogenic factors in proliferative diabetic retinopathy. Curr Diabetes Rev 2006;2:71–98.
10 Inokuchi N, Ikeda T, Imamura Y, Sotozono C, Kinoshita S, Uchihori Y, Nakamura K: Vitreous levels of insulin-like growth factor-I in patients with proliferative diabetic retinopathy. Curr Eye Res 2001; 23:368–371.
11 Hernández C, Carrasco E, García-Arumí J, Maria Segura R, Simó R: Intravitreous levels of hepatocyte growth factor/scatter factor and vascular cell adhesion molecule-1 in the vitreous fluid of diabetic patients with proliferative retinopathy. Diabetes Metab 2004;30:341–346.
12 Sieczkiewicz GJ, Herman IM: TGF-beta 1 signaling controls retinal pericyte contractile protein expression. Microvasc Res 2003;66:190–196.
13 Mintz-Hittner HA, Kuffel RR Jr: Intravitreal injection of bevacizumab (avastin) for treatment of stage 3 retinopathy of prematurity in zone I or posterior zone II. Retina 2008;28:831–838.
14 Chung EJ, Kim JH, Ahn HS, Koh HJ: Combination of laser photocoagulation and intravitreal bevacizumab (Avastin) for aggressive zone I retinopathy of prematurity. Graefes Arch Clin Exp Ophthalmol 2007;245:1727–1730.
15 Honda S, Hirabayashi H, Tsukahara Y, Negi A: Acute contraction of the proliferative membrane after an intravitreal injection of bevacizumab for advanced retinopathy of prematurity. Graefes Arch Clin Exp Ophthalmol 2008;246:1061–1063.
16 Arevalo JF, Maia M, Flynn HW Jr, Saravia M, Avery RL, Wu L, Eid Farah M, Pieramici DJ, Berrocal MH, Sanchez JG: Tractional retinal detachment following intravitreal bevacizumab (Avastin) in patients with severe proliferative diabetic retinopathy. Br J Ophthalmol 2008;92:213–216.
17 Tranos P, Gemenetzi M, Papandroudis A, Chrisafis C, Papadakos D: Progression of diabetic tractional retinal detachment following single injection of intravitreal Avastin. Eye 2008;22:862.
18 Boulton M, Gregor Z, McLeod D, et al: Intravitreal growth factors in proliferative diabetic retinopathy. Br J Ophthalmol 1997;81:228–233.
19 Kato M, Yoshimura S, Kokuzawa J, et al: Hepatocyte growth factor promotes neuronal differentiation of neural stem cells derived from embryonic stem cells. Regener Transplant 2004;15:5–8.

Kimberly A. Drenser, MD, PhD
3535 West 13 Mile Road, Suite 344
Royal Oak, MI 48073 (USA)
Tel. +1 248 288 2869, Fax +1 248 288 5644, E-Mail kimber@pol.net

Author Index

Subject Index